The Dark Energy Paradigm

Competing Interests The author has no competing interests to declare that are relevant to the content of this manuscript.

Contents

1 Dark Energy Universe 1
  1.1   Introduction 1
  1.2   Dark Energy: Different Perspectives 4
      1.2.1   The Quantum Vacuum 4
      1.2.2   Stochastic Electrodynamics 5
      1.2.3   Wheeler's Approach 6
      1.2.4   Large Number Considerations 7
  1.3   Cosmological Constant 8
  1.4   The Different Shades of Dark Energy 12
      1.4.1   The Old Ideas of Dark Energy 12
      1.4.2   Gravitation 13
      1.4.3   The New Dark Energy: From the Planck Scale
            to the Compton Scale 14
  1.5   The Origin of Inertial Mass 15
  1.6   Zero Point Energy (ZPE) 22
  1.7   Rotation Curves 24
  1.8   Trouble in Paradise? 25
  1.9   The Zero Point Energy 26
  1.10 A Cosmological Signature 28
  1.11 Ramifications of Dark Energy 29
  References 30

2 Violation of Lorentz Symmetry 35
  2.1   Introduction 35
  2.2   Modification of the Compton Scattering, the GZK Cutoff
       and so on 35
  2.3   An Extension of GZK Considerations 37
  2.4   Some Effects Due to Neutrino Behaviour 38
  2.5   Spectral Lags 38
  2.6   Spacetime Geometry and Superluminosity 39
  2.7   A Second-Order Effect 41

2.8   Back from the Future 43
2.9   The Spooky Universe 43
References 47

3   The Enigmatic Neutrino 49
3.1   Introduction 49
3.2   A Background Neutrino Model for the Cosmological
        Constant 49
3.3   Neutrino Mass 52
3.4   Neutrino Mass Measurements 53
3.5   Some Salient Aspects of Neutrinos 53
3.6   A Model for Neutrinos 56
        3.6.1   Mass, Charge, Chirality, and Other Considerations 56
        3.6.2   Two Dimensionality 56
        3.6.3   Two-Dimensional Neutrinos 60
3.7   Electrons and Neutrinos 60
3.8   Spin of the Neutrino 62
3.9   Additional Comments: Early Derivation
        of the Kerr–Newman Metric 63
3.10   Transmutations of Particles 67
3.11   Neutrino *Waves* 67
3.12   Is the Neutrino a Hybrid Particle? 70
References 73

4   The Bizarre Spacetime 77
4.1   Why Noncommutative Geometry Puts a Bound
        on Velocities? 77
4.2   Modified Klein–Gordon Equation 78
4.3   The Modified Transformations for High-Energy
        and Low-Energy Scenarios 81
4.4   The Parameter ζ 84
4.5   The Lorentz Factor Inherent in Noncommutativity 87
4.6   Minkowski Spacetime in the Light of Noncommutative
        Geometry of Modern Quantum Gravity Approaches 88
4.7   Dark Energy and Spacetime Geometry 88
4.8   Going Beyond the Standard Model* 93
4.9   Gamma Matrices and Bilinear Covariants 94
4.10   Noncommutativity and Relativity 95
        4.10.1   Noncommutativity and Boundedness of Velocity 96
        4.10.2   Composition of Velocities and Relativity 98
        4.10.3   Relativistic Mass 99
4.11   Special Relativity Derivations 102
4.12   Some Issues in Noncommutative Spacetime Geometry 104
4.13   2-D Crystals 105
References 106

5 Mystery of the Missing Dark Matter 109
    5.1   Non-Dark Matter Approaches 109
    5.2   How Essential Is Dark Matter? 110
    5.3   Dark Matter Anomalies 113
    5.4   Dark Matter: Confusion and Mystery 113
    5.5   Anomaly of Dark Matter 115
    5.6   The Expanding Universe 116
    5.7   Alternative to the Dark Matter Paradigm 118
    References 119

6 Low-Dimensional Structures 121
    6.1   Quantum Mechanical Black Holes: Low-Dimensional
          Structures 121
    6.2   One- and Two-Dimensional Behaviour 122
    6.3   Fermions and Bosons 124
    6.4   Anomalous Behaviour of Bosons and Fermions 124
    6.5   Graphene and Elementary Particles 125
    6.6   Noncommutative Repercussions 129
    6.7   Two-Dimensional Considerations 131
    6.8   Electromagnetic Effects Due to Graphene 133
    6.9   The Fractional Quantum Hall Effect in Relation
          with Graphene 134
    6.10 The Conductivity Mystery with Graphene 135
    6.11 Magnetic Effect of Noncommutativity Again 136
    6.12 The Honey-Comb Structure of Two-Dimensional Graphene 137
    6.13 Graphene and Entanglement 139
    References 139

7 A Fifth Force in Nature? 143
    7.1   Introduction 143
    7.2   Theoretical Justification for the Existence of a Fifth Force 144
    7.3   Observational Conclusions 149
    References 150

8 A Random Walk Through Miscellany 153
    8.1   Fission with a Difference 153
    8.2   Ultra-High-Energy Decaying Fermions 155
          8.2.1   The High-Energy Equation 155
          8.2.2   Possible Consequences at High Energies ... 157
    8.3   A Zeeman-Like Effect 160
    8.4   Epilogue: The Gravitation Frontier 161
    References 161

List of Figures

Fig. 1.1 This image is the detailed, all-sky picture of the infant universe created from 7 years of WMAP data. The image reveals 13.7-billion-year-old temperature fluctuations (shown as colour differences) that correspond to the seeds that grew to become the galaxies. (Picture credit: NASA) 2

Fig. 1.2 A mysterious extra-loud radio noise permeates the cosmos, preventing astronomers from observing infrared light from the first stars. The balloon-borne ARCADE instrument discovered this cosmic static on its July 2006 flight. The surprising find was that the noise was six times louder than expected. Astronomers have no clue why. (Picture credit: NASA/ARCADE) 11

Fig. 1.3 Galactic rotation curves 24

Fig. 1.4 Illustration of the Casimir effect 25

Fig. 2.1 The periscope thought (gedanken) experiment 39

Fig. 2.2 Time line diagram. (I) birth of photons 1 and 2. (II) detection of photon 1. (III) birth of photons 3 and 4. (IV) Bell projection of photons 2 and 3. (V) detection of photon 4 (picture courtesy [17]) 46

Chapter 1
Dark Energy Universe

1.1 Introduction

By 1990 it appeared that major problems of cosmology had been sorted out. Scientists were convinced that the universe began with a huge explosion, the well-known "big bang". This initial explosion was considered to be more powerful than a trillion trillion trillion trillion Hydrogen bombs exploding in unison which would fling the matter out in all directions. However physicists had hypothesized, starting in the 1930s the existence of what was called "dark matter", which would slow down the initial explosion and bring the expanding universe to a halt. This in turn would trigger a huge implosion and so on the script went. It must be mentioned that to date the exact identity of this magical dark matter is debatable. On the other hand, the dark matter hypothesis seemed to explain features like velocity curves of stars in galaxies. The point is that there was a distinct flattening of the galactic rotation curves, as if some unseen matter was pulling the stars inwards was observed. There have been many speculations about the identity of this dark matter without a definitive conclusion. For example, would they be made of supersymmetric particles, or could they be the so-called brown dwarf or Weakly Interacting Massive Particles (WIMPS) and so on. In fact, Prof. Abdus Salam speculated some two decades ago [2] "And now we come upon the question of dark matter which is one of the open problems of cosmology"....."This is a problem which was speculated upon by Zwicky in the 30s. He demonstrated that visible matter in the Coma cluster of galaxies was not sufficient to bind the galactic cluster. Oort pointed out that at least three times the mass of existing stars would be required to keep our galaxy stable". This became a central problem and focus of old cosmology $\cdots$ One would be compelled to pose the question: what exactly is dark matter? The Wilkinson Microwave Anisotropy Probe (WMAP) space mission based on 9 years of observation of the cosmic microwave

---

Zero point energy is the same as the commonly used vacuum fluctuations.

© The Author(s), under exclusive license to Springer Nature Singapore Pte Ltd. 2025
B. G. Sidharth, *The Dark Energy Paradigm*,
https://doi.org/10.1007/978-981-96-3745-4_1

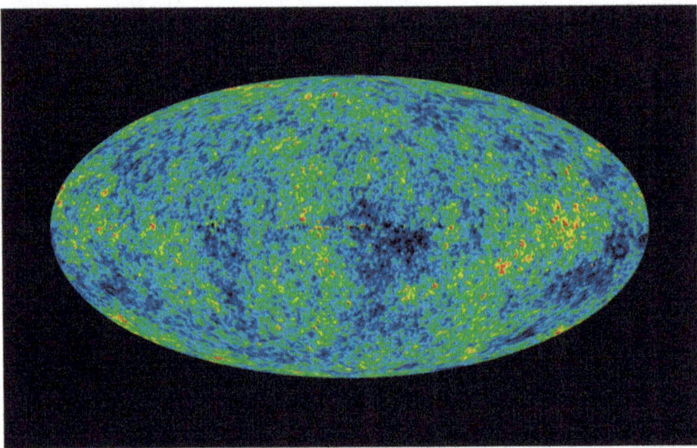

Fig. 1.1 This image is the detailed, all-sky picture of the infant universe created from 7 years of WMAP data. The image reveals 13.7-billion-year-old temperature fluctuations (shown as colour differences) that correspond to the seeds that grew to become the galaxies. (Picture credit: NASA)

background concluded the following: WMAP's measurements found that the universe is 13.7 billion years old and consists of

5% matter,

23% dark matter,

72% dark energy (see Fig. 1.1).

The universe in this picture showed the existence of enough of the mysterious dark matter to bring to a halt the expansion and eventually bring about the next collapse. That is, the universe would continue expansion up to a certain point and after that would collapse.

However, very recently, based on the observations of the Hubble Space Telescope and James Webb Space Telescope, it is being suspected that different parts of the universe could be accelerating at different rates depending on where the observation is taken [3].

Of course many other problems of subtler nature still needed to be answered. The well-known horizon problem [4] being one of them. In other words, the universe is born out of an uncontrolled and uncoordinated explosion making the big bang a random event. Because of this, different portions of the universe were flung in different directions, without any connection. So much so, light would not have had enough time to travel from one part of the universe to another. This contradicts the observed fact that the universe is on the whole consistent and homogeneous. This would give rise to a speculation of some form of faster than light intercommunication. This of course would violate the special theory of relativity.

Another conundrum arose, space should be curved due to the huge amounts of matter in it, surprisingly the universe seemed to be flat! There were other problems as well. For example, the fact that monopoles have not been found after decades of search, as they should have been.

Inflationary Cosmology

Some of these problems were sought to be explained by what has been called inflationary cosmology whereby, early on, just after the big bang the explosion was superfast [5].

What would happen in this case is that different parts of the universe, which could not be accessible by light, would now get connected. At the same time, the superfast expansion in the initial stages would smoothen out any distortion or curvature effects in space, leading to a flat universe and in the process also eliminate the monopoles.

Nevertheless, inflation theory has its problems. It does not seem to explain the cosmological constant observed since. Further, this theory seems to imply that the fluctuations it produces should continue to indefinite distances. Observation seems to imply the contrary.

A Lumpy Universe

One other feature that has been studied in detail over the past few decades is that of structure formation in the universe. To put it simply, why is the universe not a uniform spread of matter and radiation? On the contrary, it is very lumpy with planets, stars, galaxies, and so on, with a lot of space separating these objects. This has been explained in terms of fluctuations in density, that is, accidentally more matter being present in a given region. Gravitation would then draw in even more matter and so on. These fluctuations would also cause the cosmic background radiation to be non-uniform or anisotropic. Such anisotropies are in fact being observed. But this is not the end of the story. The galaxies seem to be arranged along two-dimensional structures and filaments with huge separating voids.

Suddenly in 1997 the author put forward a contra view of the universe: not so much dark matter, but what these days is called dark energy permeated the universe and dominated it. Under its influence the universe would not slow down as has been pictured earlier but rather it would accelerate, albeit very slowly [6–8]. This startling effect was observed the very next year in 1998 itself.

Perlmutter, Schmidt, and Reiss [9] got the Nobel Prize for this pathbreaking observation in 2011. This elicited some rueful comments:

A very reputed publication quoted Nobel Laureate Antony Leggett, "It is of course clear that your equation predicts an exponential (inflation-type) expansion of the current universe, hence acceleration. And it would have been nice if the Nobel committee had mentioned this". Prof. Leggett also noted the author's later work on signals of dark energy, ...looks very interesting. Prof Antony Hewish, a British astronomer who had won the Nobel for physics in 1974, also felt that the author's work should have been recognized. "You must feel gratified that your ideas in 1997 were spot on", Hewish wrote.

"..I would guess that while you did not share the prize this year, the fact that you predicted something that was the key to the others getting the prize should make your chances quite high for the future", wrote another Nobel Laureate Douglas Osheroff. He noted, "I certainly do appreciate that you are one of the very few to have recognized, on theoretical grounds, the possible need to reintroduce a non-zero cosmological constant ahead of the supernova experiments!"

Nobel Laureate Prof. I. Prigogine described the work as "Very interesting" and, "I agree with you that spacetime has a stochastic underpinning". And so on.

1.2 Dark Energy: Different Perspectives

1.2.1 The Quantum Vacuum

It is commonly accepted that at absolute zero degree Kelvin there is no motion whatsoever: when thermodynamic motion ceases this is expected. In practice this is not found to be so as was observed by Nernst, the discoverer of the third law of thermodynamics. The superfluidity of Helium is caused by quantum mechanics—rather than the third law. We are referring to the spooky motion of supercooled Helium. Let us explore the quantum vacuum and the related zero point energy (ZPE).

The Zero Point Energy and Its Manifestations

- A familiar effect, the Casimir effect shows the existence of a force between gold plates without charge placed in a charge-free medium. This effect experimentally confirms the presence of the elusive zero point energy or quantum vacuum energy.
- A miniscule vibration of an electron which orbits the nucleus of an atom can be observed by the Lamb shift. Leading to the conclusion that the zero point energy is causing this vibration.
- Then there is the question of the anomalous quantum mechanical gyromagnetic ratio $g = 2$ leading to the quantum mechanical spin half and so on [10–12].

The quantum vacuum replaces the placid aether of yore. Now, electrons and positrons are incessantly created and destroyed in this turbulent medium, practically instantaneously. Of course the limits are set by the Heisenberg uncertainty principle to circumvent energy conservation violation. Quantum vacuum maybe looked upon as another state of matter, maybe a compromise between being and nonbeing. The Vedic seers, while referring to the universe, succinctly put it, thousands of years ago as: "Neither existence, nor non-existence".

The lowest state of any quantum field with zero momentum and energy maybe treated as quantum vacuum. Heisenberg's principle attributes an infinite value to

ZPE and has to be "renormalized", implying that it is to be ignored. Unlike with the older concept of aether, quantum vacuum can take on different properties. Quantum vacuum is responsible for effects like quark confinement, which, roughly speaking, implies that it would not be possible to isolate an independent or free quark. Quantum vacuum also explains the spontaneous breaking of symmetry of the electroweak theory put forward by Abdus Salaam and others, as well as vacuum polarization. In vacuum polarization, electron-like particles are surrounded by a cloud of other oppositely charged particles. This tends to reduce or mask the main charge and so on. Regions having similarity with domain structures of ferromagnets could be found due to vacuum fluctuations. All elementary electron-magnets are aligned with their spins in a particular direction in a ferromagnet. But some regions with the spins aligned differently could also be found.

Zeldovich and others as also the author [13, 14] emphasized that such a quantum vacuum can cause cosmic repulsion. The problem with this approach is the huge value of the cosmological constant that it throws up. In fact the universe would blow up almost immediately after it is created, with such a value. This is the so-called cosmological constant problem emphasized by Weinberg [15].

1.2.2 Stochastic Electrodynamics

There is another approach, wherein stochastic electrodynamics treats the ZPE as independent and primary and attributes to it quantum mechanical effects [16]. It may be re-emphasized that the ZPE results in the well-known experimentally verified Casimir effect [17, 18]. We would also like to point out that contrary to popular belief, the concept of aether has survived over the decades through the works of Dirac, Vigier, Prigogine, string theorists like Wilzeck, and others [19–21]. It appears that even Einstein himself continued to believe in this concept.

The argument proceeds as follows: Elementary particles are formed as a result of the fluctuations of energy in the electromagnetic field. Einstein believed this to be true. According to Wilzeck, Einstein looked for a formulation where particles and radiation could have a unified origin. For Einstein, the fields were primary. Later on he tried to find precisely such a unified formulation. He was not successful in his efforts.

We will now argue that indeed this can happen. In the words of Wheeler [12], "From the zero point fluctuations of a single oscillator to the fluctuations of the electromagnetic field to geometrodynamic fluctuations is a natural order of progression..."

1.2.3   Wheeler's Approach

Let us consider Wheeler's approach. The starting point of this approach is a harmonic oscillator in its ground state with a probability amplitude:

$$\psi(x) = \left(\frac{m\omega}{\pi\hbar}\right)^{1/4} e^{-(m\omega/2\hbar)x^2},$$

where x is the displacement from its position of classical equilibrium. The fluctuation of the oscillator occurs in an interval given by

$$\Delta x \sim (\hbar/m\omega)^{1/2}.$$

The electromagnetic field is comprised of an infinite set of independent oscillators, bearing amplitudes X_1, X_2, etc. The oscillators with amplitudes X_1, $X_2 \cdots$ have a probability which now becomes the product of individual oscillator amplitudes:

$$\psi(X_1, X_2, \cdots) = exp[-(X_1^2 + X_2^2 + \cdots)].$$

This would have to have a suitable normalization factor. This expression for probability gives the probability amplitude ψ for a magnetic field with configuration $B(x, y, z)$. This maybe be given by the Fourier coefficients X_1, X_2, $\cdots$ or directly in terms of the magnetic field configuration itself by

$$\psi(B(x, y, z))$$
$$= P \, exp\left(-\int\int \frac{\mathbf{B}(x_1) \cdot \mathbf{B}(x_2)}{16\pi^3 \hbar c r_{12}^2} d^3x_1 d^3x_2\right)$$

P being a normalization factor. Let us now specialize to the case of the magnetic field which is everywhere zero except in a region which has a dimension l. In this region, the magnetic field is of the order of $\sim \Delta B$. The probability amplitude with the above configuration is proportional to

$$\exp\left[-\left((\Delta B)^2 l^4/\hbar c\right)\right].$$

This implies that the fluctuational energy in a volume of length l will be [12, 22, 23]

$$B^2 \sim \frac{\hbar c}{l}. \tag{1.1}$$

It maybe be noted that in Eq. (1.1) if l were to be taken to be the Compton wavelength of a typical elementary particle, then its energy mc^2 would be recovered. This can be verified easily. It was noted earlier by the author, Rueda [1, 24] and others, that inertial mass and energy may be deduced based on viscous resistance to the ZPE. This may also be deduced from quantum mechanical considerations, restricted to the Compton scale. This brings us back to the result in the context of the ZPE. The inverse dependence of the length scale and the energy (or momentum) is shown by (1.1).

According to Einstein, such a condensation from the background electromagnetic field yielded the elementary particles (Cf. [14, 25] for details). The above result follows along the same lines. In the sequel, we also take the pion to represent a typical elementary particle, as in the literature.

1.2.4 Large Number Considerations

To proceed, as there are $N \sim 10^{80}$ such particles in the universe, we get, consistently,

$$Nm = M, \tag{1.2}$$

where M is the mass of the universe and m is taken to be the pion mass as it represents a typical elementary particle, as mentioned earlier. It must be remembered that the energy of gravitational interaction between the particles is very much insignificant compared to the above electromagnetic considerations.

In the following, we will use N as the sole cosmological parameter.

We next invoke the well-known relation [26–28]

$$R \approx \frac{GM}{c^2}, \tag{1.3}$$

where M can be obtained from (1.2). We can arrive at (1.3) in different ways. For example, in a uniformly expanding Friedman universe, we have

$$\dot{R}^2 = 8\pi G \rho R^2 / 3.$$

In the above if we substitute $\dot{R} = c$ at R, the radius of the universe, we get (1.3).

We now use the fact that given N particles, the (Gaussian) fluctuation in the particle number is of the order $\sqrt{N}$ [6, 8, 28–31], while a typical time interval for the fluctuations is $\sim \hbar/mc^2$, the Compton time, the fuzzy interval we encountered within which there is no meaningful physics. So particles are created and destroyed but the ultimate result is that $\sqrt{N}$ particles are created just as this is the net displacement in a random walk of unit step. So we have

$$\frac{dN}{dt} = \frac{\sqrt{N}}{\tau}, \tag{1.4}$$

whence on integration we get (remembering that we are almost in the continuum region, that is, $\tau \sim 10^{-23}\,\text{s} \approx 0$)

$$T = \frac{\hbar}{mc^2}\sqrt{N}. \tag{1.5}$$

We can easily verify that Eq. (1.5) is indeed satisfied where T is the age of the universe. Next by differentiating (1.3) with respect to t we get

$$\frac{dR}{dt} \approx HR, \tag{1.6}$$

where H in (1.6) can be identified with the Hubble constant, and on using (1.3) H is given by

$$H = \frac{Gm^3c}{\hbar^2}. \tag{1.7}$$

Equations (1.2), (1.3), and (1.5) show that in this formulation, the correct mass, radius, Hubble constant, and age of the universe can be deduced given N, the number of particles, as the sole cosmological or large-scale parameter. We observe that at this stage we are not invoking any particular dynamics—the expansion is due to the random creation of particles from the ZPE background. Equation (1.7) can be written as

$$m \approx \left(\frac{H\hbar^2}{Gc}\right)^{\frac{1}{3}}. \tag{1.8}$$

Equation (1.8) has been empirically known as an "accidental" or a "mysterious" relation. As observed by Weinberg [32], this is unexplained: it relates a single cosmological parameter H to constants from microphysics. We will touch upon this micro–macro nexus again. In our formulation, Eq. (1.8) is no longer a mysterious coincidence but rather a consequence of the theory. From Eq. (1.8), we can deduce that the value of $m \approx 10^{-25}$ gm which is the pion mass. Indeed in the large number theory, as is well known, and as mentioned earlier, the pion is considered to be a typical elementary particle as it takes part in various interactions including hadronic.

1.3 Cosmological Constant

As (1.7) and (1.6) are not exact equations but rather, order of magnitude relations, it follows, on differentiating (1.6) that a small cosmological constant Λ is allowed such that

$$\Lambda \leq O(H^2).$$

This is consistent with observation and shows that Λ is very small—this has been a puzzle, the so-called cosmological constant problem alluded to, because in conventional theory, it turns out to be huge and unacceptable [33]. But it poses no problem in this formulation. This is because of the characterization of the ZPE as independent and primary in our formulation, this being the mysterious dark energy. We shall further characterize Λ later in this chapter.

To proceed we observe that because of the fluctuation of $\sim \sqrt{N}$ (due to the ZPE), there is an excess electrical potential energy of the electron, which in fact we identify as its inertial energy. That is [8, 28],

$$\sqrt{N} e^2 / R \approx mc^2.$$

On using (1.3) in the above, we recover the well-known gravitation-electromagnetism ratio, viz.:

$$e^2 / Gm^2 \sim \sqrt{N} \approx 10^{40} \tag{1.9}$$

or without using (1.3), we get, instead, the well-known so-called Weyl–Eddington formula:

$$R = \sqrt{N} l. \tag{1.10}$$

It appears that (1.10) was first noticed by H. Weyl. In fact (1.10) is the spatial counterpart of (1.5). If we combine (1.10) and (1.3), we get

$$\frac{Gm}{lc^2} = \frac{1}{\sqrt{N}} \propto T^{-1}, \tag{1.11}$$

where in (1.11), we have used (1.5). Following Dirac we treat G as the variable, rather than the quantities m, l, c and $\hbar$ which we will call microphysical constants because of their central role in atomic (and sub-atomic) physics.

Next if we use G from (1.11) in (1.7), we can see that

$$H = \frac{c}{l} \frac{1}{\sqrt{N}}. \tag{1.12}$$

Thus apart from the fact that H has the same inverse time dependence on T as G, (1.12) shows that given the microphysical constants, and N, we can deduce the Hubble constant also, as from (1.12) or (1.7).

Using (1.2) and (1.3), we can now deduce that

$$\rho \approx \frac{m}{l^3} \frac{1}{\sqrt{N}}. \tag{1.13}$$

Next (1.10) and (1.5) give

$$R = cT. \tag{1.14}$$

Equations (1.13) and (1.14) are consistent with observation.

Finally, we observe that using M, G and H from the above, we get

$$M = \frac{c^3}{GH}.$$

This relation is required in the Friedman model of the expanding universe (and the steady-state model too). In fact if we use in this relation, the expression

$$H = c/R$$

which follows from (1.12) and (1.10), then we recover (1.3).

As we saw the above model predicts a dark-energy-driven ever-expanding and accelerating universe with a small cosmological constant while the density keeps decreasing. Moreover, mysterious large number relations like (1.7), (1.13), or (1.10) which were considered to be miraculous accidents now follow from the underlying theory. This seemed to go against the accepted idea that the density of the universe equaled the critical density required for closure and that aided by dark matter, the universe was decelerating. However, as noted, from 1998 onwards, following the work of Perlmutter, Schmidt, and co-workers, these otherwise apparently heretic conclusions have been vindicated.

It may be mentioned that the observational evidence for an accelerating universe was the American Association for Advancement of Science's Breakthrough of the Year, 1998 while the evidence for nearly 75 percent of the universe being dark energy, based on the Wilkinson Microwave Anisotropy Probe (WMAP) and the Sloan Sky Digital Survey was the Breakthrough of the Year, 2003 [34, 35]. See Fig. 1.1.

However one issue which has not yet been settled is the fundamental question, "What exactly is dark energy?" In the author's analysis, to re-emphasize, this was a mysterious energy called the zero point energy which had been known for a long time, though its connection with dark energy had not been thought of. One way of understanding this is: if an object be suspended in deep space, with no forces whatever acting on it, then according to the usual laws of physics, it would not budge even a little. However the mysterious zero point energy would be buffeting it and contrary to conventional wisdom, it would be vibrating. It leads to what is called the cosmological constant, a repulsive force which Einstein had in a different and mistaken context introduced nearly a century earlier and soon retracted.

Other scholars have tried to characterize dark energy with different and novel descriptions and interpretations, some of them very complicated and some of them invoking string theory. However, it is not clear if these models are fruitful. If the author's characterization of dark energy as the zero point energy is correct, then it should leave a cosmic footprint, namely, a background of radio waves (including microwaves) [36, 37]. The problem is that the earth's ionosphere filters out these radio waves. However, in the recent few years, NASA has conducted balloon-based experiments called the ARCADE experiments.

These balloons reach a height of some 40 km, just outside the ionosphere. They lead to a shocking discovery which was called the space roar by the experimenters.

This was a persistent hiss of radio waves. It is known that radio waves are received by the Earth from objects like Quasars or other radio sources. But the special feature of the space roar is that it does not come from any specific source or collection of sources. It is a uniform background. The balloon-borne instrument named ARCADE stands for the Absolute Radiometer for Cosmology, Astrophysics, and Diffuse Emission. In July 2006, the instrument was launched from NASA's Columbia Scientific Balloon Facility in Palestine, Texas, and flew to an altitude of 120,000 feet, where the atmosphere thins into the vacuum of space (see Fig. 1.2). Indeed the author's 1997 model and subsequent work shows that there should be exactly such a cosmic radio wave background as a signature of the all-pervading zero point energy. This is therefore a vindication of the 1997 model.

Fig. 1.2 A mysterious extra-loud radio noise permeates the cosmos, preventing astronomers from observing infrared light from the first stars. The balloon-borne ARCADE instrument discovered this cosmic static on its July 2006 flight. The surprising find was that the noise was six times louder than expected. Astronomers have no clue why. (Picture credit: NASA/ARCADE)

Finally, it may be asked: Why did this treatment of zero point energy work and not other formulations? The argument can be summarized as: (i) the Compton wavelength rather than the Planck length (or points) is the feature here [38] and (ii) fluctuations (Gaussian) are factored in.

In fact the picture that emerges is starting from the background of the quantum vacuum, which essentially is a sea of Planck-sized or Planck-scale oscillators, there would be a phase transition at the Hagedorn temperature leading to the production of pions [37]. These pions in turn would form groups of other particles via the QCD interactions (Cf. Ref. [37]) as discussed in the above reference. What is remarkable is that this approach throws up a formula for the mass spectrum which covers all non-leptonic elementary particles, in most cases with an error of $1-1.5\%$ (Cf. also [39]).

1.4 The Different Shades of Dark Energy

1.4.1 The Old Ideas of Dark Energy

What is today called dark energy has been around since the early part of the last century itself. It was discussed by Nernst, the father of the third law of Thermodynamics, in the context of superfluidity. He believed that this phenomenon was caused by dark energy, which for him was the zero point energy. He further believed that the universe is in an ocean of dark energy and that particles condense out of such an ocean [40]. Yet these ideas seem to have led nowhere and they died a death. The idea of the zero point energy in the context of the cosmological constant was revived some decades later, thanks to the work of Zeldovich and others. But again these ideas didn't seem to lead anywhere. On the contrary, they led to what Weinberg called the cosmological constant problem [33]. What would happen is that the universe would blow up in almost no time. So strong would be the repulsive force.

So in the late 90s it was believed that the universe was slowing down, thanks to the preponderance of dark matter which ostensibly comprised 95% of the universe. In fact, this was the standard big bang model of the 90s. As noted earlier in 1997, the author put forward the model of a dark-energy-driven accelerating universe, though with a small cosmological constant [6, 8]. Though there was initial scepticism, finally the dark-energy-driven accelerating universe was accepted [1, 7, 14].

Let us briefly analyse this matter—in particular, what was the difference between the earlier ideas and the author's 1997 model. In the earlier case, dark energy was conceived in the context of the Planck scale so that the cosmological constant or the vacuum energy density would be huge, some 10^{120} times its observed values. This is because the cosmological constant (or vacuum energy density) would be

$$\Lambda \sim 0\left(\frac{1}{l^4}\right) \tag{1.15}$$

which would be much too high if l were the Planck length $\sim 10^{-33}$ cm. This would mean, as noticed, that the universe would blow up almost as soon as it was born.

On the other hand, this would be the "scale of gravitation". Let us see why.

1.4.2 Gravitation

Cercignani [41] had used quantum oscillations, though just before the dark energy era—these were the usual earlier zero point oscillations. Invoking gravitation, what he proved was, in his own words, "Because of the equivalence of mass and energy, we can estimate that this (i.e. chaotic oscillations) will occur when the former will be of the order of

$$G[\hbar\omega)c^{-2}]^2[\omega^{-1}c]^{-1} = G\hbar^2\omega^3 c^{-5}, \tag{1.16}$$

where G is the constant of gravitational attraction and we have used as distance, the wavelength. This must be less than the typical electromagnetic energy $\hbar\omega$. Hence (from (1.16)),

$$\omega < (G\hbar)^{-1/2} \cdot c^{5/2} \tag{1.17}$$

which gives a gravitational cutoff for the frequency in the zero-point energy". In other words, he deduced that there has to be a maximum frequency of oscillators given by

$$G\hbar\omega_{max}^2 = c^5 \tag{1.18}$$

for the very existence of coherent oscillations (and so a coherent universe). We would like to point out that if we use the above in Eq. (1.18) we get the well-known relation

$$Gm_P^2 \approx \hbar c \tag{1.19}$$

m_P being the Planck mass. Let us see how this happens: In (1.18), we use the fact that

$$\hbar\omega = m_P c^2,$$

after rewriting it as

$$G\hbar^2\omega^2 = c^5 \cdot \hbar$$

This leads to

$$Gm_P^2 c^4 = \hbar c^5,$$

where (1.19) follows. This shows that at the Planck scale the gravitational and electromagnetic strengths are of the same order. This is not surprising because it was the very basis of Cercignani's derivation—if indeed the gravitational energy is greater than that given in (1.19) that is greater than the electromagnetic energy, then the zero point oscillators would become chaotic and incoherent—there would be no physics.

However all this refers to a classical description because we are working here at the Planck scale. In fact (1.18) and (1.19) can be alternatively deduced, considering these Planck oscillations as phonons (Cf. Ref. [7]). So the picture that emerges is the following: till frequencies with a cutoff at the Planck length, a classical description of the zero point energy is valid. This is the domain of gravitation. Beyond that up to the Compton scale we have a quantum mechanical description and this leads to electromagnetism and other interactions. In any case, it is the zero point energy or dark energy all the way.

1.4.3 The New Dark Energy: From the Planck Scale to the Compton Scale

In the author's model, it was not the Planck scale, but rather the Compton scale. This would sort out all the problems but there still has to be a mechanism for transiting from the Planck scale to the Compton scale. This can be achieved in a few ways. The first is by considering oscillations in dark energy at the Planck scale, what may be called Planck oscillations. It has been shown in great detail that when these Planck oscillations become coherent, the situation can be modelled in terms of Bénard cells in a liquid at a phase transition. We would then end up at the Compton scale [42].

Yet another way of looking at this would be via loop quantum gravity considerations.

Loop Quantum Gravity Considerations

Another theory of quantum gravity, namely, loop quantum gravity (LQG) is also an attempt at reconciling quantum mechanics with general relativity. It is based directly on Einstein's geometric formulation of general relativity. As a theory, LQG hypothesizes that the structure of space and time consists of finite loops intermeshed into a fine network. Such networks of loops are called spin networks. The evolution of a spin network, or spin foam, has a scale, interestingly above the order of a Planck length, approximately 10^{-35} m, (somewhat like this author's work) and smaller scales are considered meaningless. In postulating this, not just matter, but space itself would have an atomic structure.

Loop quantum gravity provides a description of the microstructure of quantum physical space. This physical space is characterized by a Planck scale discreteness. This discreteness emerges in a natural manner from quantum theory and provides a mathematical realization of Wheeler's intuition of a spacetime "foam" [43].

There is also another way to reach the same conclusion: The argument is completely different though ultimately all methods are referring to the same phenomena.

We start with Wheeler's quantum foam, that is, Planck mass "point" particles, presumably created in a big bang event. The extension of these particles is given by the Planck length, while the mass is 10^{-5} gms.

Another approach would be by considering a Herman Weyl type of reconciliation which was rejected originally by Einstein as being ad hoc. However, in the author's approach, this aspect is completely different. We must remember that the Dirac matrices are really bi-spinors describing positive-energy solutions say ϕ and negative energy solutions χ. These two have different behaviours under reflection with $\phi \mapsto \phi$ and $\chi \mapsto -\chi$. Once this is recognized we can recover the Weyl-like equations reconciling gravitation and electromagnetism bypassing Einstein's original objection. This matter has been discussed in great detail in the author's book, The Thermodynamic Universe [1].

1.5   The Origin of Inertial Mass

References [44, 45]. Our starting point is an equation deduced by Feynman [46, Chap. 8] in a simple way,

$$\imath\hbar\frac{\partial C(x)}{\partial t} = \frac{-\hbar^2}{2\,m'}\frac{\partial^2\,C(x)}{\partial x^2}, \tag{1.20}$$

where $C(x) \equiv |\psi(x)\rangle$ is the probability amplitude for the particle to be at the point x at some given moment of time.

To deduce Eq. (1.20), we follow the development of [46, Chap. 8] and define a complete set of base states by the subscript $\imath$ and $U(t_2, t_1)$ the time elapse operator that denotes the passage of time between instants t_1 and t_2, t_2 greater than t_1. We denote by $C_\imath(t) \equiv \langle\imath|\psi(t)\rangle$, the amplitude for the state $|\psi(t)\rangle$ to be in the state $|\imath\rangle$ at time t, and

$$\langle\imath|U|j\rangle \equiv U\imath j, \quad U\imath j(t + \Delta t, t) \equiv \delta_{\imath j} - \frac{\imath}{\hbar}H_{\imath j}(t)\Delta t.$$

We can now deduce from the superposition of states principle that

$$C_\imath(t + \Delta t) - \sum_j[\delta_{\imath j} - \frac{\imath}{\hbar}H_{\imath j}(t)\Delta t]C_j(t)$$

and finally, in the limit,

$$\imath\hbar\frac{dC_\imath(t)}{dt} = \sum_j H_{\imath j}(t)C_j(t), \tag{1.21}$$

where the matrix $H_{ij}(t)$ is identified with the Hamiltonian operator. To facilitate comparison we stick to the notation and development as given in [46]. Before proceeding to derive the Schrodinger equation, we apply Eq. (1.21) to the simple case of a two-state system ($i, j = 1, 2$), respectively (cf. Ref. [46]). This will provide a physical picture for the later work. For a two-state system, we have

$$\imath\hbar\frac{dC_1}{dt} = H_{11}C_1 + H_{12}C_2$$

$$\imath\hbar\frac{dC_2}{dt} = H_{21}C_1 + H_{22}C_2$$

leading to two stationary states of energies $E - A$ and $E + A$, where $E \equiv H_{11} = H_{22}$, $A = H_{12} = H_{21}$. We can choose our zero of energy such that $E = 2A$. Indeed as has been pointed out by Feynman, when this consideration is applied to the hydrogen molecular ion, the fact that the electron has amplitudes C_1 and C_2 of being with either of the hydrogen atoms manifests itself as an attractive force which binds the ion together, with an energy of the order of magnitude $A = H_{12}$.

To proceed, we consider in (1.21) the $\imath$ to be the space point $x_\imath$ and we denote $C(x_n) \equiv C_n$ the probability amplitude for the particle to be at this space point. Further let $x_{n+1} - x_n = b$. Then considering only the point x_n and its neighbours $x_{n\pm1}$, Eq. (1.21) goes over into

$$\imath\hbar\frac{\partial C(x_n)}{\partial t} = EC(x_n) - AC(x_n - b) - AC(x_n + b). \qquad (1.22)$$

In the limit $b \to 0$, with our choice of the arbitrary zero of energy, (1.22) goes over into Eq. (1.20) where we have now dropped the subscript distinguishing the space point and $m' = \hbar^2/2Ab^2$.

We now observe that while Eq. (1.20) resembles the free Schrodinger equation, as has been pointed out by Feynman, m' is not really the inertial mass, but an "effective mass" that emerges from the probability amplitude for the particle to be found at a neighbouring point. So (1.20) is not the Schrodinger equation.

The Schrodinger equation can be obtained from (1.20) if it can be shown that m' can somehow be replaced by m, the inertial mass. This is what we propose to do.

To start with, let us suppose that the particle has no mass other than the effective mass m', so that we can treat Eq. (1.20) as the Schrodinger-type equation for such a particle which has only amplitude to be at neighbouring points. Let us now suppose that the particle acquires non-zero probability amplitude to be present non-locally at other than neighbouring points. We can then no longer work with Eqs. (1.22) and (1.20). We will have to use the full Eq. (1.21) which explicitly exhibits this possibility. We rewrite Eq. (1.21) as

$$\imath\hbar\frac{dC_\imath(t)}{dt} = H_{\imath\imath}C_\imath(t) + H_{\imath,\imath-1}C_{\imath-1}(t) + H_{\imath,\imath+1}C_{\imath+1}(t) + \sum_j H_{\imath,\imath+j}(t)C_j(t), (j = \pm2, \pm3,)$$

or as in the transition of Eqs. (1.22)–(1.20),

$$\imath\hbar\frac{\partial C(x)}{\partial t} = \frac{-\hbar^2}{2\,m'}\frac{\partial^2\,C(x)}{\partial x^2} + \int H(x, x')C(x')dx', \qquad (1.23)$$

where we have replaced H_{ij} by $H(x, x')$ and the points x_i are in the limit taken for the time being to be a continuum. This is as in the well-known case of the non-local Schrodinger equation for a non-local potential [47] but for a particle having only an effective mass.

The matrix $H(x, x')$ gives the probability amplitude for the particle at x to be found at x', that is,

$$H(x, x') = \langle\psi(x')|\psi(x)\rangle, \qquad (1.24)$$

where as usual we write $C(x) \equiv \psi(x)(\equiv |\psi(x)\rangle)$, the state of a particle at the point x.

Usually the amplitude $H(x, x')$ is non-zero only for neighbouring points x and x', that is, $H(x, x') = f(x)\delta(x - x')$. But if $H(x, x')$ is not of this form, then there is a non-zero amplitude for the particle to "jump" to an other than neighbouring point. In this case, $H(x, x')$ may be described as a non-local amplitude. Indeed such non-local amplitudes are implicit in the Dirac equation also and this will be commented on.

We now give a quick derivation of how the inertial mass emerges from Eq. (1.23). The non-local Schrodinger equation (1.23), given only the effective mass m', can be written, with the help of (1.24), as

$$\imath\hbar\frac{\partial\psi}{\partial t} = \frac{-\hbar^2}{2\,m'}\frac{\partial^2\psi}{\partial x^2} + \int \psi^*(x')\psi(x)\psi(x')U(x')dx', \qquad (1.25)$$

where
(i) $U(x) = 1$ for $|x| < R$, R arbitrarily large and also $U(x)$ falls off rapidly as $|x| \to \infty$; $U(x)$ has been introduced merely to ensure the convergence of the integral; and
(ii) $H(x, x') = \langle\psi(x')\psi(x)\rangle - \psi^*(x')\psi(x)$.

Equation (1.25) is an integro-differential equation of degree three.

The presence of the, what at first sight may seem troublesome, non-linear and non-local term, viz. the last term on the right side of (1.25) can be satisfactorily explained.

In (1.25), in the first approximation $\psi(x)$ can be taken to be the solution of the Schrodinger-like Eq. (1.20), viz.

$$\imath\hbar\frac{\partial\psi}{\partial t} = \frac{-\hbar^2}{2\,m'}\frac{\partial^2\psi}{\partial x^2}. \qquad (1.26)$$

In effect, we linearize (1.25), so that we get

$$\imath\hbar\frac{\partial\psi}{\partial t} = \left[-\frac{\hbar^2}{2\,m'}\frac{\partial^2}{\partial x^2} + m_0\right]\psi, \qquad (1.27)$$

where

$$m_0 = \int \psi^*(x')\psi(x')U(x')dx'.$$

In operator language, (1.27) becomes

$$\overline{H} = \frac{p^2}{2\,m'} + m_0, \tag{1.28}$$

where $\overline{H}$ is the Hamiltonian operator, $\vec{p}$ is the momentum operator, p is magnitude, $p^2 = \vec{p} \cdot \vec{p}$ and where, what can now be anticipated as a rest mass like term m_0, appears for a particle assumed not to have any rest mass in the absence of the non-local amplitude term in (1.25). Also we have replaced the Hamiltonian matrix H by $\overline{H}$ to stress that, to start with, in (1.23) and (1.25), the particle has no inertial mass. To facilitate comparison with the usual theory, we next multiply both sides of (1.28) by the constant $\frac{m'}{m}$, where

$$m = (m_0 m')^{\frac{1}{2}}/c,$$

c being the velocity of light (the reason for the appearance of the velocity of light c can be seen below (cf. Eq. (1.30))) and the constant could be absorbed into the state vector, whose direction is all that matters. We then get

$$\widehat{H} = \frac{p^2}{2\,m} + mc^2. \tag{1.29}$$

The physical meaning of (1.29) is now clear. In an expansion of the classical relativistic expression for energy,

$$E = (p^2 c^2 + m^2 c^4)^{1/2}$$

as is well known, if we keep terms up to the order $(p/mc)^2$, we get

$$E = \frac{p^2}{2\,m} + mc^2. \tag{1.30}$$

We can now easily identify m in (1.29) with the rest mass on comparing this equation with (1.30). (Interestingly it is not accidental that Eq. (1.29) corresponds to the approximation (1.30) as will be seen below.) If further, we denote

$$H = \hat{H} - mc^2,$$

where H can be easily identified with the usual kinetic energy operator (or energy operator in non-relativistic theory, remembering that we are considering a free particle only), (1.29) becomes

$$H = \frac{p^2}{2\,m}.\tag{1.31}$$

In a strictly non-relativistic context, where the rest energy of the particle is not considered, the Hamiltonian is given by (1.31); otherwise, it is given approximately by (1.29). We get from (1.31), the Schrodinger equation:

$$\imath\hbar\frac{\partial\psi}{\partial t} = -\frac{\hbar^2}{2\,m}\frac{\partial^2\psi}{\partial x^2}.\tag{1.32}$$

All these considerations can be considered in a postulative development [44] and also generalized in a simple way to three dimensions, but as there is no new physical insight, the details are not given.

The physical origin of the rest mass is clear from Eq. (1.29): in the two-state hydrogen molecular ion case considered earlier, it was the amplitude for the single electron to be with one hydrogen atom or the other which showed up as a binding energy. Similarly the amplitude of a particle to be at x or x', viz. the second term on the right side of Eq. (1.25) manifests itself as an (attractive) energy, which may be called the mass energy of the particle or the self-energy or the energy of self-interaction. This can be seen to be the particle's inertial mass.

We now come to the non-local term in Eq. (1.25), the term which gives the inertial mass. Non-locality implies superluminal velocities and the breakdown of causality which is not permissible in general. However without any contradiction to the theory it is well known that quantum mechanics allows such non-locality, owing to the uncertainity principle [32], within the Compton wavelength of a particle. So there is no contradiction if the non-local integral in (1.25) is taken within the region of the particle's Compton wavelength, that is, the inertial mass is a result of non-local processes within the Compton wavelength of the particle.

Indeed the usual Dirac equation also has a non-local character: The operator $c\vec{\alpha}.\vec{p} + \beta mc^2$ is equivalent to and replaces the non-local square-root operator $(-\hbar^2\nabla^2 + m^2c^4)^{1/2}$. Here also the non-local effects in the form of negative energies are encountered—again within the Compton wavelength region (cf. Ref. [48]).

In the light of the preceding considerations, we can derive the Schrodinger equation from an alternative angle: it appears that the "point" particle is really spread over the non-locality region $\sim \bar{b} = \frac{\hbar}{mc}$, the Compton wavelength. Further, the energy of the particle, i.e. the energy tied up within this region, viz. 2A is the inertial mass energy mc^2. We could now speak of the amplitude for the particle at x to be found (locally) at a neighbouring point $x + b$, except that in the limit, $b \to \bar{b}$ (and not as earlier 0). The effective mass m' in Eq. (1.20) is then given by

$$m' = \frac{\hbar}{2\Lambda b^2} = m,$$

that is the mass itself!

So, Eq. (1.20) can be interpreted as the Schrodinger equation.

It is worth re-emphasizing that it is the force of binding of non-local positions within the Compton wavelength, rather like the hydrogen molecular ion binding, that manifests itself as inertial mass.

Finally, we briefly comment on the appearance of the extra mass energy term in equations like (1.23), (1.25), (1.28), (1.29), or (1.30) [44, 49, 50].

The Schrodinger equation is really the limiting case of the Dirac equation in which process an inessential phase factor is dropped. Another way of looking at this is that the constant potential $m_o c^2$ does not affect the dynamics. That is the reason why the Schrodinger equation is not Galilean invariant, as a non-relativistic theory should be, and in fact exhibits the Sagnac effect, which a strictly Galilean invariant theory should not [51].

The convergence of the above formulation and the Bohm hydrodynamical formulation is evident once we restrict ourselves to the Compton wavelength and luminal velocities. The particle is now a relativistic fluid vortex circulating along a ring of radius equal to the Compton wavelength. For more details on the Bohm formulation, see [44] and references therein.

Let us now consider distances of the order of the Compton wavelength. At this level, quantum mechanical phenomena like *zitterbewegung*, negative energy solutions, and luminal velocities come into play. Taking a route through relativistic vortices, monopoles, and classical considerations, we will lead to the model of leptons and quarks as what may be called "Quantum Mechanical Kerr–Newman Black Holes" (QMKNBH), wherein features of quantum mechanics and general relativity are inextricably inter-woven.

If ψ is the wave function in the quantum foam then the probability amplitude that such a particle would be at a point x is given by as deduced by the author in 1996 [44] and details of the transition from the discrete to the continuous case are briefly outlined earlier.

$$
\iota \hbar \frac{\partial \psi}{\partial t} = \frac{-\hbar^2}{2\, m'} \frac{\partial^2 \psi}{\partial x^2}
$$
$$
+ \int \psi^*(x')\psi(x)\psi(x')U(x')dx'. \tag{1.33}
$$

The integral is over a small δ interval around the point x while $U(x)$ has been inserted for convenience: It is 1 in this interval and 0 outside. This immediately leads to the Landau–Ginzburg equation:

$$
-\frac{\hbar^2}{2\, m} \nabla^2 \psi + \beta |\psi|^2 \psi = -\alpha \psi. \tag{1.34}
$$

At this stage we would like to point out that the justification for treating the variables in Eq. (1.34) as spanning a continuum is as follows: firstly as we are dealing with miniscule intervals of the order of the Compton scale, Eq. (1.34) holds to a very good approximation. More mathematically this can be seen from Wheeler's space-time foam where lengths below the Planck length have no meaning. As discussed in detail and as Dirac pointed out, our concept of spacetime or space begins after averaging over intervals of the Planck or Compton scales.[1]

In this case, there is a coherence length given by

$$\xi = \frac{\hbar v_F}{\Delta} \tag{1.35}$$

Δ being the energy mc^2, where m is the mass of the particle in the δ interval and $v_F = c$ is the maximal velocity.

To see how this happens, we invoke the Landau–Ginzburg theory of superconductivity. In this case, as is well known, we have the equation which resembles (1.34), in the absence of electromagnetism. Specializing to the case of a homogeneous superconductor, we get

$$\alpha\psi + \gamma|\psi|^2\psi + \frac{1}{2m}(-\iota\hbar\nabla - 2eA)^2\psi = 0, \tag{1.36}$$

where

$$\alpha\psi + \gamma|\psi|^2\psi = 0$$

in the first approximation. A solution can immediately be seen to be

$$|\psi|^2 = -\alpha/\gamma$$

$$|\psi|^2 = \frac{\alpha(T - T_c)}{\gamma}. \tag{1.37}$$

All this is exactly as in the Landau–Ginzburg theory. The point about (1.37) is that as we approach the critical temperature, i.e. $T \to T_c$ from below, $|\psi|^2 \to 0$ everywhere.

This means that in our case, the wave function vanishes. Furthermore, in this theory, above the critical phase, we have the coherence length given by (1.35).

This can clearly be seen to be the Compton wavelength as v_F is c. In fact, Wigner and Salecker have argued at length that there can be no physical measurements within the Compton scale [52]. Furthermore, as is known, the interesting aspects of the critical point theory ([53]) are universality and scale. Broadly, this means that diverse physical phenomena follow the same route at the critical point, on the one hand, and on the other, this can happen at different scales, as exemplified for example, by the course graining techniques of the Renormalization Group [54].

So not only do we come up to the Compton scale and hence the 1997 cosmology but also we are able to deduce the so-called Large Number formulae of Eddington, Weinberg, and others which have been a puzzle for over a century.

---

[1] As noted earlier, this conclusion can also be obtained by the use of LQG.

Let us see how this can happen. Equation (1.35) is the starting point. To highlight this point we note that in critical point phenomena we have the reduced order parameter $\bar{Q}$ (which gives the fraction of the excess of new states) and the reduced correlation length $\bar{\xi}$ (which follows from (1.35)). Near the critical point we have well-known relations [55] like

$$(\bar{Q}) = |t|^{\beta}, (\bar{\xi}) = |t|^{-\nu}$$

whence

$$\bar{Q}^{\nu} = \bar{\xi}^{\beta}. \tag{1.38}$$

In (1.38) typically $\nu \approx 2\beta$. As $\bar{Q} \sim \frac{1}{\sqrt{N}}$ because $\sqrt{N}$ particles are created fluctuationally, or in the transition given N particles, and in view of the fractal two dimensionality of any path as explained by Abbott and Weiss [56]

$$\bar{Q} \sim \frac{1}{\sqrt{N}}, \bar{\xi} = (l/R)^{2}. \tag{1.39}$$

This gives back the Eddington formula:

$$R = \sqrt{N}l.$$

This is of course the spread R in a random walk of step length l and consisting of N steps. Other relations can also be deduced without much difficulty (Cf. Ref. [1]).

It may be mentioned that Beck and Meckay too have considered a phase transition though a few years later which throws up an ultimate energy of $1.7Tz$ [38] though it is not clear what exactly this represents.

Recently the problems with the old cosmological constant and dark energy are being articulated again. For example, Harry Cliff of the LHC and Cambridge University comments [57] that physics is coming to a dead end because of this.

1.6   Zero Point Energy (ZPE)

The question which arises is, what exactly is this dark energy? In the author's 1997 formulation, it was the zero point energy (ZPE). What is quite remarkable is that to date, it remains the same. This ZPE is ubiquitous and generally has been renormalized.

Let us look at this from another point of view. We observe that the coherent N' Planck oscillators referred to above could be considered to be a degenerate Bose assembly. In this case as is well known we have

$$v = \frac{V}{N'}$$

(Cf. Ref. [29] here z of the usual theory ≈ 1). V the volume of the universe $\sim 10^{84}\,\mathrm{cm}^3$. Whence

$$v = \frac{V}{N'} \sim 10^{-36}.$$

So that the wavelength

$$\lambda \sim (v)^{1/3} \sim 10^{-12}\,\mathrm{cm} = l. \tag{1.40}$$

What is very interesting is that (1.40) gives us the Compton length of a typical elementary particle like the pion. So from the Planck oscillators we are able to recover the elementary particles exactly as before [58, 59].

So our description of the universe at the Planck scale is that of an entangled wave function as in

$$\psi = \sum_n c_n \phi_n. \tag{1.41}$$

However, we perceive the universe at the elementary particle or Compton scale, where the random phases would have weakened the entanglement, and we have the description as in

$$\psi = \sum_n b_n \bar{\phi}_n. \tag{1.42}$$

Does this mean that N elementary particles in the universe are totally incoherent in which case we do not have any justification for treating them to be in the same spacetime?

We can argue that they still interact among each other though in comparison this is "weak". For instance, let us consider the background ZPE whose spectral frequency is given by

$$\rho(\omega) = \mathrm{const} \cdot \omega^3. \tag{1.43}$$

Whence from

$$(\Delta B)^2 \geq \hbar c / L^4 \tag{1.44}$$

the energy in the entire volume $\sim L^3$ is given by

$$\Delta E \sim \hbar c / L^2.$$

As we are in the Compton region, where $L = cT$ and furthermore $1/T$ is the frequency ω, we are led to the well-known Eq. (1.43).

If there are two particles at A and B separated by a distance r, then those wavelengths of the ZPE which are at least $\sim r$ would connect or link the two particles. Whence the force of interaction between the two particles is given by, remembering that $\omega \propto 1/r$,

$$\text{Force} \propto \int_r^\infty \omega^3 dr \propto \frac{1}{r^2}. \tag{1.45}$$

Thus from (1.45) we are able to recover the familiar Coulomb law of interaction. The background ZPE thus enables us to recover the action at a distance formulation and also the three dimensionality of space. In fact, a similar argument has been given by other authors to recover from QED the Coulomb law—here the carriers of the force are the virtual photons, that is, photons whose life time is within the Compton time of uncertainty permitted by the Heisenberg uncertainty principle.

It is thus possible to synthesize the field and action at a distance concepts, once it is recognized that there is the ZPE and there are minimum spacetime intervals at the Compton scale [60]. Many of the supposed contradictions arise because of our characterization in terms of spacetime points and a differentiable manifold. Once the minimum cutoff at the Planck scale is introduced, this leads to the physical Compton scale and a unified formulation free of divergence problems.

We now make a few comments.

1.7 Rotation Curves

In the intervening years, since there have been interesting developments. For example, cosmologist Riess has observed that there is roughly a 7% increase in the acceleration of the universe. This has been described and explained by the author as being further evidence of the existence of dark energy [61]. Further the author has been arguing that the role of dark matter is much exaggerated. There is very strong data of early galactic rotation curves [62] (see Fig. 1.3).

In fact, Miligrom [63, 64] has argued that a slight modification of Newtonian mechanics can explain away dark matter. This theory is also called the *Mond* theory. The author, on the other hand, and a few others have pointed out the ad hoc nature of this fix. The author himself prefers a continuous but slow change of the gravitational constant with time, somewhat on the lines of the "Dirac large number theory" [32, 65]. The author points out in [1, 7, 14, 66] in detail his distributional effect on G (cf. also [67]).

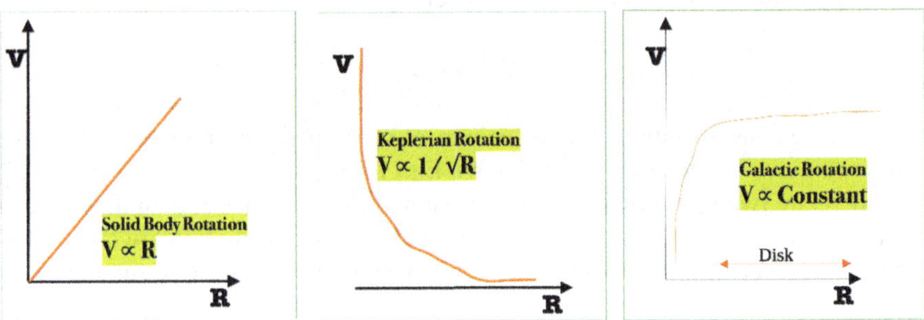

Fig. 1.3 Galactic rotation curves

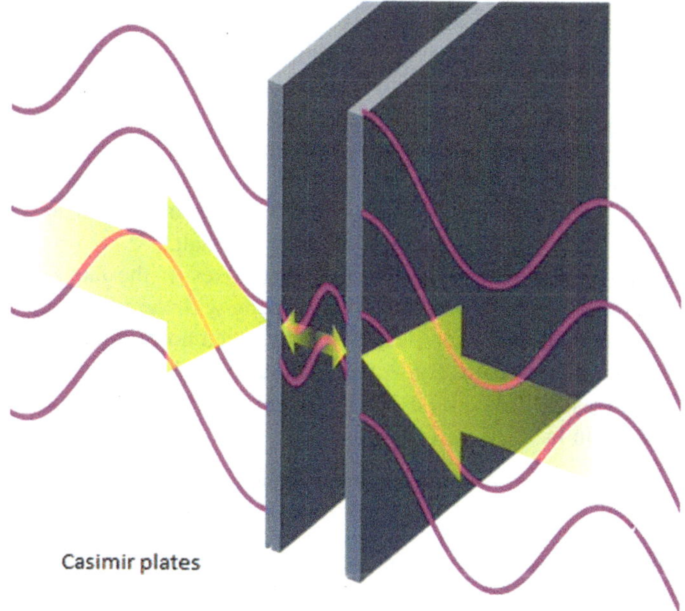

Fig. 1.4 Illustration of the Casimir effect

Casimir Effect

Subsequently it has been felt that the original Casimir experiment (See Fig. 1.4) can be explained by usual quantum effects rather than by the vacuum energy hypothesis (cf. [68]).

Furthermore, it turns out, as explained in detail in [1], vacuum energy can be related to the inverse square force and three dimensions as we have just seen. But perhaps the most surprising finding has been that dark energy leads to Lorentz invariance [68]. This is because of the dependence of the spectrum on ω^3 as we saw in Eq. (1.43).

In retrospect it looks quite amazing that the zero point energy (ZPE) at the microscopic level leads to the macroscopic Lorentz invariance.

It is well known that the Casimir effect can be demonstrated through the gold leaf microscope. Here we have energy conservation. But in Sidharth's Dark energy model, there is no energy conservation and in this case, the Gold leaf microscope demonstrates the conversion of dark energy into real energy.

1.8   Trouble in Paradise?

Recently, Nobel Laureate Adam Reiss and co-workers have reported an anomalous finding: Under the present cosmological model, roughly 70% of the content of the universe is dark energy, some 25% is dark matter and about 5% is ordinary matter. Furthermore, the universe is not only expanding, but accelerating in the process, as pointed out, driven by dark energy, though the acceleration or cosmological constant seems to be small. As we noted earlier, the latest observations report that the acceleration is 8% more than the present cosmological constant model admits.

It will be recalled that in 1997 the accepted cosmological model was that of a dark matter-dominated universe with less than 5% of visible matter. This universe would be expanding, though decelerating before it came to a halt.

Is this a similar situation is the question that arises, if the latest observations are correct. In other words is there something wrong with the present cosmological model? It may be mentioned that to date there has been no definite sighting of dark matter, which therefore remains conjectural. On the other hand, in the author's work [1, 7], dark matter is replaced by a gravitational constant G which decreases very slowly with time. In fact

$$\left| \frac{\dot{G}}{G} \right| \leq 10^{-11}/yr. \tag{1.46}$$

For example, we could consider an interaction of dark energy with a fermionic field, contained in dark matter, these fermions being neutrinos [69, 70]. Attempts have been made to formulate an equation of state for a dark energy fluid [71]. Questions have also been asked whether we have dissipative cosmology or conservative cosmology as a result [72], while a generalized second law has also been studied [73]. The coincidence problem is also being studied, viz. why the energy density of dark energy is roughly of the same order as a cosmological critical density [74–76]. The author himself suggested earlier a model based on background cosmic neutrinos [67, 77]. Even more recently the problem has been studied by Li et al. who have compared nine different popular models for dark energy [78]. The models under consideration are the cosmological constant model, two equation of state parameterization models, the generalized Chaplygin gas model, two Dvali–Gabadadze–Porrati models, and three holographic dark energy models. All of these models are well-known dark energy candidates and have attracted considerable attention in the past without leading anywhere in particular.

1.9 The Zero Point Energy

We first observe that the concept of a zero point energy (ZPE) or quantum vacuum (or vacuum energy) is an idea whose origin can be traced back to Max Planck himself. Quantum field theory attributes the ZPE to the virtual quantum effects of an already present electromagnetic field [79].

In a very intuitive and preliminary way, Faraday could conceive of magnetic effects in vacuum in connection with his experiments on induction. Based on this, an aether was used for the propagation of electromagnetic waves in Maxwell's theory of electromagnetism, which in fact laid the stage for special relativity. This aether was a homogenous, invariable, non-intrusive, material medium which could be used as an absolute frame of reference, at least for certain chosen observers. However, the experiments of Michelson and Morley towards the end of the nineteenth century lead to its downfall, and thus was born Einstein's special theory of relativity in which there is no such absolute frame of reference.

Very shortly thereafter the advent of quantum mechanics lead to its rebirth in a new and unexpected avatar [80]. Essentially there were two new ingredients in what is today called the quantum vacuum. The first was a realization that classical physics had allowed an assumption to slip in unnoticed: in a source or charge free "vacuum", one solution of Maxwell's equations of electromagnetic radiation is no doubt the zero solution. But there is also a more realistic non-zero solution. That is, the electromagnetic radiation does not necessarily vanish in empty space.

The second ingredient was the mysterious prescription of quantum mechanics, the Heisenberg uncertainty principle, according to which it would be impossible to precisely assign momentum and energy, on the one hand, and spacetime location, on the other. Clearly the location of a vacuum with no energy or momentum cannot be specified in spacetime.

This leads to what is called a zero point energy. For instance, a harmonic oscillator, a swinging pendulum, for example, according to classical ideas has zero energy and momentum in its lowest position. But the Heisenberg uncertainty endows it with a fluctuating energy. This fact was recognized by Einstein himself way back in 1913 who, contrary to popular belief, retained the concept of aether though from a different perspective [81]. It also provides an understanding of the fluctuating electromagnetic field in vacuum.

This mysterious zero point energy or quantum vacuum energy has since been experimentally confirmed in effects like the Casimir effect which demonstrates a force between uncharged parallel plates separated by a charge-free medium, the Lamb shift which demonstrates a minute oscillation of an electron orbiting the nucleus in an atom—as if it was being buffetted by the zero point energy—the anomalous quantum mechanical gyromagnetic ratio $g = 2$ and so on [11, 12, 82, 83].

The quantum vacuum is a violent medium in which charged particles like electrons and positrons are constantly being created and destroyed, almost instantly, within the limits permitted by the Heisenberg uncertainty principle for the violation of energy conservation.

There are also claims that the virtual photons of the quantum vacuum have been realized as real photons, in an endorsement of the dynamical Casimir effect (Cf. Ref. [84]). One might call the quantum vacuum as a new state of matter, a compromise between something and nothingness.

The quantum vacuum can be considered to be the lowest state of any quantum field, having zero momentum and zero energy. The properties of the quantum vacuum can under certain conditions be altered, which was not the case with the erstwhile aether. In modern particle physics, the quantum vacuum is responsible for phenomena like quark confinement, a property whereby it would be impossible to observe an independent or free quark, the spontaneous breaking of symmetry of the electroweak theory, vacuum polarization wherein charges like electrons are surrounded by a cloud of other oppositely charged particles tending to mask the main charge, and so on. There could be regions of vacuum fluctuations comparable to the domain structures of ferromagnets. In a ferromagnet, all elementary electron-magnets are aligned with their spins in a certain direction. However, as noted earlier, there could be special regions wherein the spins are aligned differently.

There is another approach, sometimes called stochastic electrodynamics, which treats the ZPE as primary and attributes to it quantum mechanical effects [16].

We would next like to observe that the energy of the fluctuations in the background electromagnetic field could lead to the formation of elementary particles. Indeed this was Einstein's belief. As he observed as early as 1920 itself [85], "... according to our present conceptions, the elementary particles are... but condensations of the electromagnetic field".

In the words of Wilzeck, [81], "Einstein was not satisfied with the dualism. He wanted to regard the fields, or aethers, as primary. In his later work, he tried to find a unified field theory, in which electrons (and of course protons, and all other particles) would emerge as solutions in which energy was especially concentrated, perhaps as singularities. But his efforts in this direction did not lead to any tangible success". We will return to this point later.

1.10   A Cosmological Signature

Let us now refer to an interesting experiment performed by NASA: The ARCADE 2 experiment. This is the second-generation Absolute Radiometer for Cosmology, Astrophysics, and Diffuse Emission (ARCADE 2) instrument. It comprises a balloon-borne experiment. This takes measurements of the radiometric temperature of the cosmic microwave background and galactic and extra-galactic emission, performed by A. Kogut and others [86]. It discovered cosmic radio noise that is six times louder than what we would expect from old ideas.

Another intriguing cosmological footprint of dark energy [48, 87] maybe inferred. As dark energy is the all pervading ZPE, and we bear in mind that the ZPE causes the Lamb Shift (as well as via *zitterbewegung*). In the hydrogen atom, the Lamb shift is $\sim 1000\,\text{MHz}$ or about 30 cm wavelength, corresponding to the radio region. The several dissipative processes in space would lead one to expect that the ZPE would lead to isotropic radio waves, which are not related to any particular radio source, say like quasars, in the sky. Such sources would be expected to have a wavelength of 30 cm or more. As the ionosphere reflects radio waves coming from outer space, such radio waves are difficult to detect on the earth. This is where the ARCADE 2 experiment was of immense help.

1.11 Ramifications of Dark Energy

1. We would like to reiterate that the zero point energy at the microscopic level leads to Lorentz invariance as also the three dimensionality of space.
2. With regard to the varying G question, a good review has been given (see Uzan [88]).
3. Very recently a team from Portsmouth University [89] has made the most detailed study yet of millions of galaxies and have come to the conclusion that the universe is flat!
4. There have been attempts by a number of physicists to tap dark energy. But we must bear in mind that dark energy is more like the superfluidity of hydrogen—it doesn't follow ordinary mechanics. This is brought out by the author's thought experiment, as pointed out: the dark energy pendulum [90]. It is almost as if the pendulum bob is executing a three-dimensional random walk.
5. It should be clear that dark energy disrupts the laws of thermodynamics, starting with the very first law and the second law, viz. the entropy law. However, it must be mentioned that Nernst, the father of the third law of thermodynamics himself believed in a scheme, not dissimilar to the scheme of the author.
6. Something similar was done by Prof. Ilya Prigogine in his ideas [91]. He considered stochastic models and steady-state cosmology, out of which particles are irreversibly created by instability or fluctuation. In his words (Cf. Ref. [91]) "The big bang was an event associated with an instability within the medium that produced our universe. Although our universe has an age, the medium that produced our universe has none... We consider the big bang an irreversible process par excellence from a pre universe that we call quantum vacuum. This eventually would result from an instability in the pre universe..."
[92].
7. From the above large number relations, we can also conclude that

$$\frac{Gm}{lc^2} = \frac{1}{\sqrt{N}}$$

This is the same as

$$e^2/Gm^2 \sim \sqrt{N} \approx 10^{40}.$$

This gives the correct ratio of the field strengths of electromagnetism and gravitation. But it also shows the nature of gravitation as a distributed force.
8. One may wonder why there is, what maybe called a micro–macro divide. The microuniverse is generally the universe we refer to, but the macrouniverse is more like the universe at large. It is remarkable that in spite of this, there is a nexus between the two.
9. From the ethereal to the real:
We had already deduced that the mysterious dark energy or ZPE, surprisingly follows what maybe called a shadow electromagnetic field. There is a shadow

electromagnetic field tensor, current vector, etc. As noted, this field is ubiquitous. More importantly, as it is spread everywhere, though weak, we can exploit this electromagnetic field—it is a matter of technology.

10. An interesting point to note is that, normally special relativity is a purely classical concept, whereas from the above discussion it appears that it is quantum mechanical, depending on probability amplitudes.

References

1. B.G. Sidharth, *The Thermodynamic Universe* (World Scientific, Singapore, 2008)
2. B.G. Sidharth, Encounter with Abdus Salaam. New Adv. Phys. **1**(1), 1–15 (1997)
3. A.G. Riess et al., JWST observations reject unrecognized crowding of cepheid photometry as an explanation for the hubble tension at 8σ confidence. Astrophys. J. Lett. **962**(1), L17 (2024)
4. A. Zee, *Unity of Forces in the Universe*, vol. 2 (World Scientific, 1982)
5. A.D. Linde, A new inflationary universe scenario: a possible solution of the horizon, flatness, homogeneity, isotropy and primordial monopole problems. Phys. Lett. B **108**(6), 389–393 (1982)
6. T. Piran, R. Ruffini, *The Eighth Marcel Grossmann Meeting* (World Scientific, Singapore, 1999)
7. B.G. Sidharth, *The Universe of Fluctuations* (Springer, 2005)
8. B.G. Sidharth, The universe of fluctuations. Int. J. Mod. Phys. A **13**(15), 2599–2612 (1998)
9. S. Perlmutter, Discovery of a supernova explosion at half the age of the Universe. Nature **391**(6662), 51–54 (1998)
10. P.W. Milonni. *The Quantum Vacuum: An Introduction To Quantum Electrodynamics* (Academic, 2013)
11. R. Podolny, *Something Called Nothing* (Mir Publishers, 1986)
12. C.W. Misner, K.S. Thorne, *and J* (Freeman, A. Wheeler. Gravitation, 2000)
13. Y.B. Zel'dovich. Pis'ma zh. eksp. teor. fiz. 6 883 Zel'dovich Ya B 1967. JETP Lett. **6**, 316 (1967)
14. B.G. Sidharth, *Chaotic Universe* (Nova Science, New York, 2001)
15. S. Weinberg, The cosmological constant problem. Rev. Mod. Phys. **61**, 1–23 (1989). (Jan)
16. E. Santos, Stochastic electrodynamics and the Bell inequalities, in *Open Questions in Quantum Physics* (Springer, 1985), pp. 283–296
17. V.M. Mostepanenko, N.N. Trunov, The Casimir effect and its applications. Phys. Usp. **31**(11), 965–987 (1988)
18. S.K. Lamoreaux, Demonstration of the Casimir force in the 0.6 to $6\mu m$ range. Phys. Rev. Lett. **78**, 5–8 (1997)
19. V. Hushwater, Repulsive Casimir force as a result of vacuum radiation pressure. Am. J. Phys. **65**(5), 381–384 (1997)
20. N.C. Petroni, J.P. Vigier, Dirac's aether in relativistic quantum mechanics. Found. Phys. **13**(2), 253–286 (1983)
21. T.D. Lee, Statistical Mechanics of Quarks and Hadrons (North-Holland Publishing Company, 1981), pp. 237–249
22. B.G. Sidharth, Fluctuations in the quantum vacuum. Chaos, Solitons & Fractals **14**(1), 167–169 (2002)
23. B.G. Sidharth, Quantum mechanics from cosmic fluctuations. Chaos, Solitons & Fractals **15**(1), 25–27 (2003)
24. A. Rueda, B. Haisch, Inertia as reaction of the vacuum to accelerated motion. Phys. Lett. A **240**(3), 115–126 (1998)
25. A. Marlow, *Quantum Theory and Gravitation* (Elsevier, 2012)

26. B.G. Sidharth, The cosmology of fluctuations. Chaos, Solitons & Fractals **16**(4), 613–620 (2003)
27. L. Nottale, *Fractal Spacetime and Microphysics: Towards a Theory of Scale Relativity* (World Scientific, 1993)
28. S. Hayakawa, Atomism and cosmology. Prog. Theor. Phys. Suppl. **65**, 532–541 (1965)
29. K. Huang, *Statistical Mechanics* (Wiley Eastern, New Delhi, 1975)
30. B.G. Sidharth, Large numbers and the time variation of physical constants. Int. J. Theor. Phys. **37**(4), 1307–1312 (1998)
31. B.G. Sidharth, Quantum Black Holes, in *Frontiers of Quantum Physics-1997.* ed. by C.H. Lim (Springer, Singapore, 1998), pp.308–309
32. S. Weinberg, *Gravitation and Cosmology: Principles and Applications of the General Theory of Relativity* (Wiley, 1972)
33. S. Weinberg, Baryon-and lepton-nonconserving processes. Phys. Rev. Lett. **43**(21), 1566 (1979)
34. J. Glanz, Astronomy: cosmic motion revealed. Science **282**(5397), 2156–2157 (1998). https://www.science.org/doi/abs/10.1126/science.282.5397.2156a
35. Science's Breakthrough of The Year: Illumination of the Dark, Expanding Universe. www.sciencedaily.com/releases/2003/12/031219073445.htm
36. B.G. Sidharth, S.R. Valluri, Cosmic Background Radiation. Int. J. Theor. Phys. **54**(8), 2792–2797 (2015)
37. B.G. Sidharth, Negative energy solutions and symmetries. Int. J. Mod. Phys. E **20**(10), 2177–2188 (2011)
38. C. Beck, M.C. Mackey, Electromagnetic dark energy. Int. J. Mod. Phys. D **17**(01), 71–80 (2008)
39. B.G. Sidharth, The Pion Model. Int. J. Mod. Phys. E **20**(06), 1527–1532 (2011). (Jun)
40. H. Kragh, Walther nernst: grandfather of dark energy? Astron. Geophys. **53**(1), 1.24–1.26, 02 (2012)
41. C. Cercignani, On a nonquantum derivation of Planck's distribution law. Found. Phys. Lett. **11**(2), 189–199 (1998)
42. B.G. Sidharth, The new cosmos. Chaos, Solitons & Fractals **18**(1), 197–201 (2003)
43. C. Rovelli, Loop quantum gravity. Living Rev. Relativ. **11**, 1–69 (2008)
44. B.G. Sidharth, Quantum mechanical black holes: towards a unification of quantum mechanics and general relativity. Indian J. Pure Appl. Phys. **35**, 456 ff (1997)
45. B.G. Sidharth, Quantum Mechanical Black Holes: Towards a Unification of Quantum Mechanics and General Relativity (1998)
46. R.P. Feynman, R.B. Leighton, M. Sands, *The Feynman lectures on physics*, vol. 3 (American Association of Physics Teachers, 1965)
47. H.S.W. Massey, *The Theory of Atomic Collisions* (Clarendon Press, 1949)
48. J.D. Bjorken, S.D. Drell, *Relativistic Quantum Mechanics* (Mcgraw-Hill College, 1964)
49. D. Dieks, G. Nienhuis, Relativistic aspects of nonrelativistic quantum mechanics. Am. J. Phys. **58**(7), 650–655 (1990)
50. J.A. Okolowski, M. Slomiana, Comment on "Relativistic aspects of nonrelativistic quantum mechanics," by Dennis Dieks and Gerard Nienhuis [Am. J. Phys. 58, 650–655 (1990)]. Am. J. Phys. **61**(4), 376–377 (1993)
51. J. Anandan, Sagnac effect in relativistic and nonrelativistic physics. Phys. Rev. D **24**(2), 338 (1981)
52. H. Salecker, E.P. Wigner, Quantum limitations of the measurement of space-time distances. Phys. Rev. **109**, 571–577 (1958). (Jan)
53. P. Davies, *The New Physics* (Cambridge University Press, 1992)
54. K.G. Wilson, The renormalization group and critical phenomena. Rev. Mod. Phys. **55**(3), 583 (1983)
55. B.G. Sidharth, Scaled quantum effects. Chaos, Solitons & Fractals **20**(4), 701–703 (2004)
56. L.F. Abbott, M.B. Wise, Dimension of a quantum-mechanical path. Am. J. Phys. **49**(1), 37–39 (1981)
57. H. Cliff, Have We Reached the end of Physics? https://sites.psu.edu/carlydunfordrcl/2019/10/24/have-we-reached-the-end-of-physics-ted-talk/,

58. B.G. Sidharth, Planck-scale phenomena. Found. Phys. Lett. **15**(6), 577–583 (2002)
59. B.G. Sidharth, The Planck scale in the universe. Found. Phys. Lett. **17**(5), 503–506 (2004)
60. B.G. Sidharth, Instantaneous action at a distance in a holistic universe. *Instantaneous Action at a Distance in Modern Physics:" pro" and" contra"* (1999), p. 257
61. B.G. Sidharth, *Spacetime Energy* (Nova Science, New York, 2019)
62. R. Genzel, Förster N. M. Schreiber, H. Übler, P. Lang, T. Naab, R. Bender, L.J. Tacconi, E. Wisnioski, S. Wuyts, T. Alexander, et al., Strongly baryon-dominated disk galaxies at the peak of galaxy formation ten billion years ago. Nature **543**(7645), 397–401 (2017)
63. M. Milgrom, A modification of the Newtonian dynamics-Implications for galaxies. Astrophys. J. **270**, 371–389 (1983)
64. M. Milgrom, Modified dynamics predictions agree with observations of the Hiota kinematics in faint dwarf galaxies contrary to the conclusions of Lo, Sargent, and Young. Astrophys. J. **429**, 540–544 (1994)
65. J.V. Narlikar, *An Introduction to Cosmology* (Cambridge University Press, 1993)
66. B.G. Sidharth, Effects of varying G. Nuovo Cimento della Societa Italiana di Fisica [Sezione] B **115,** 151–154 (2000)
67. B.G. Sidharth, A note on the cosmic neutrino background and the cosmological constant. Found. Phys. Lett. **19**(7), 757–759 (2006)
68. Calphys Institute. Zero Point Energy (2013). http://www.calphysics.org/zpe.html
69. A. de la Macorra, Interacting dark energy: decay into fermions. Astroparticle Phys. **28**(2), 196–204 (2007)
70. P.S. Corasaniti, M. Kunz, D. Parkinson, E.J. Copeland, B.A. Bassett, Foundations of observing Dark Energy dynamics with the Wilkinson microwave anisotropy probe. Phys. Rev. D **70**, 083006 (2004). (Oct)
71. S. Nojiri, S.D. Odintsov, The new form of the equation of state for Dark Energy fluid and accelerating universe. Phys. Lett. B **639**(3–4), 144–150 (2006)
72. M. Szydłowski, O. Hrycyna, Dissipative or conservative cosmology with Dark Energy? Ann. Phys. **322**(12), 2745–2775 (2007)
73. G. Izquierdo, D. Pavón, Dark Energy and the generalized second law. Phys. Lett. B **633**(4–5), 420–426 (2006)
74. N. Straumann, Dark Energy: recent developments. Mod. Phys. Lett. A **21**(14), 1083–1097 (2006)
75. U. França, Dark Energy, curvature, and cosmic coincidence. Phys. Lett. B **641**(5), 351–356 (2006)
76. W. Zhao, Y. Zhang, Coincidence problem in YM field Dark Energy model. Phys. Lett. B **640**(3), 69–73 (2006)
77. B.G. Sidharth, Modeling Dark Energy and the Higgs Boson. New Adv. Phys. **6**(2), 71–75 (2012)
78. M. Li, X.D. Li, X. Zhang, Comparison of Dark Energy models. Science China Physics, Mechanics and Astronomy, pp. 1631–1645
79. W. Greiner, B. Muller, J. Rafelski, *Quantum Electrodynamics of Strong Fields* (Springer, 1985)
80. B.G. Sidharth, The mysterious Dark Energy, in *Einstein and Poincaré*, ed. by Dvoeglazov, p. 79 ff. Apeiron
81. F. Wilczek, The Persistence of Ether. Phys. Today **52**(1), 11–13 (1999)
82. T.D. Lee, *Particle Physics and Introduction to Field Theory (Contemporary Concepts in Physics, Vol. 1)*, vol. 9 (1981)
83. B.G. Sidharth, The Dark Energy paradigm. AIP Conf. Proc. **1446**(1), 244–251 (2012)
84. C.M. Wilson, G. Johansson, A. Pourkabirian, et al., Observation of the dynamical Casimir effect in a superconducting circuit. Nature **479**(7373), 376–379 (2011)
85. A. Einstein, G.B. Jeffery, W. Perrett, Sidelights on relativity. Philos. Rev. **34**(2) (1925)
86. D.J. Fixsen, A. Kogut et al., Arcade 2 measurement of the absolute sky brightness at 3–90 ghz. Astrophys. J. **734**(1), 5 (2011)
87. C. Itzykson, J.B. Zuber, *Introduction to Quantum Field Theory* (McGraw-Hill International Book Company, 1980)

88. J.P. Uzan, Varying constants, gravitation and cosmology. Living Rev. Relat. **14**(1), 1–155 (2011)
89. S. Nadathur et al., Testing low-redshift cosmic acceleration with large-scale structure. Phys. Rev. Lett. **124**(22), 221301 (2020)
90. B.G. Sidharth, *When the Universe Took a U-turn* (World Scientific, 2009)
91. I. Prigogine, I. Stengers, *The End of Certainty, Rediscovering Becoming Insights from an Oriental Perspective* (The Free Press., 1996)
92. A. Grib, Particle creation and the origin of the visible universe. Gravitation & Cosmology - Gravit Cosmol **2**, 189–191, 01 (1996)

Chapter 2
Violation of Lorentz Symmetry

2.1 Introduction

Violation of Lorentz invariance has been widely researched including by the author [1, 2], and references therein. This has been supported by experimental evidence. The approach followed in this chapter is to use a modified energy–momentum relation and examine its ramifications. The modified energy–momentum relation is the so-called Snyder–Sidharth relation [3, 4]. This relation chooses the Compton length (including the Planck length) as the fundamental length. This gives a form of the Compton scattering formula as shown by the author [5]. In this approach, gamma rays travelling large distances (cosmologically speaking) undergo cumulative modified Compton lags as they collide with electrons and, in the process, develop an observable time lag. These scattering processes result in an observable time separation [6–8] of two gamma rays with different energies. We show that Lorentz symmetry violation for photons with high eV can be detected if such a simulation of high-energy rays is possible. Also using the modified energy–momentum relation an extension of the GZK cutoff maybe possible. This could explain the observations of AGASA collaboration [9] and the results of Hayashida [10]. These observations suggest some instances which violate the GZK cutoff. That is, the detection of ultra-high-energy cosmic rays (UHECR) is greater than about 10^{20} eV.

2.2 Modification of the Compton Scattering, the GZK Cutoff and so on

We begin with the Snyder–Sidharth energy–momentum relation:

$$E^2 = m^2 c^4 + p^2 c^2 - \frac{\lambda^2 l^2 c^2}{\hbar^2} p^4 \tag{2.1}$$

B. G. Sidharth, *The Dark Energy Paradigm*,
https://doi.org/10.1007/978-981-96-3745-4_2

the kinetic energy is thus

$$p_1^2 c^2 = T(T + 2mc^2), \tag{2.2}$$

where $p_1^2 = p^2 - \frac{\lambda^2 l^2}{\hbar^2} p^4$. References [11, 12] $|\lambda| \sim 10^{-3}$. As noted by Andri-ambololona and Rakotonirina, in "A Study of the Dirac-Sidharth Equation", the Compton scattering is given by [4]

$$\frac{c}{\nu'} - \frac{c}{\nu} = \frac{\hbar}{mc}(1 - \cos\theta).$$

Here, ν' is the modified frequency due to the new inputs.

Using the Hamiltonian (2.1) and taking the momentum of the recoil electron as $\vec{p}$ this formula becomes [13]

$$\frac{1}{\nu'} - \frac{1}{\nu} = \frac{\hbar}{mc^2}(1 - \cos\theta) - \frac{(2\pi)^2\lambda^2 l^2}{2mc^2\hbar^3 c^2 \nu\nu'}(\Omega^2 + 2mc^2\Omega)^2,$$

where $\Omega = \hbar\nu - \hbar\nu'$. Writing $l = \frac{\hbar}{mc}$ for the Compton length of the electron the Compton scattering formula becomes

$$\frac{1}{\nu'} - \frac{1}{\nu} = \frac{\hbar}{mc^2}(1 - \cos\theta) - \frac{(2\pi)^2\lambda^2 l^3}{2c^3\nu\nu'}[(\nu - \nu')^2 + \frac{c}{l}(\nu - \nu')]^2. \tag{2.3}$$

For the purpose of quantification, we take a simplified model consisting of two gamma rays propagating across large distances, cosmologically speaking. Let us assume that these rays are Compton scattered by an electron through the same angle θ (say). From Eq. (2.3), the difference in the time taken by the two gamma rays to be Compton scattered is

$$\begin{aligned}
\Delta t &= \left(\frac{1}{\nu} - \frac{1}{\nu_0}\right) - \left(\frac{1}{\nu_1} - \frac{1}{\nu_{01}}\right) \\
&= \frac{(2\pi)^2\lambda^2 l^3}{2c^3}\left[\frac{\{(\nu_{01} - \nu_1)^2 + \frac{c}{l}(\nu_{01} - \nu_1)\}^2}{\nu_1\nu_{01}}\right] \\
&\quad - \frac{\{(\nu_0 - \nu)^2 + \frac{c}{l}(\nu_0 - \nu)\}^2}{\nu\nu_0}
\end{aligned} \tag{2.4}$$

where ν_0 and ν and ν_{01} and ν_1 are the initial and final frequencies of the first and the second gamma rays, respectively. We assume (without loss of generality) that the energy of the second gamma ray is greater than that of the first gamma ray, therefore

$$\Delta t > 0. \tag{2.5}$$

This implies that the gamma ray with higher energy or higher frequency is faster than that which has lower energy or lower frequency, as was also seen by Pavlopoulos [14].

We can conclude now that using the Snyder–Sidharth dispersive relation (2.1) and the modified Compton scattering relation (2.3) we are able to prove that gamma rays with higher energy travel faster than those with lower energies. On the other hand, if λ is neglected,

$$\Delta t = 0,$$

that is, there is no time lag. But, observations show that there is a time lag between higher energy and lower energy gamma rays. We can conclude that Eqs. (2.1) and (2.3) are meaningful.

Thus we can infer that several successive Compton scattering processes can cause time lags. It may also be stated that if we consider two Compton scatterings of energies ≥ 1 GeV (gamma rays and high-energy gamma rays) then we may observe a prominent time lag and the violation of Lorentz invariance.

2.3 An Extension of GZK Considerations

If a particle has total energy E and momentum $\vec{p}$, then its four momentum is

$$\mathbf{p} = \begin{pmatrix} E/c \\ \vec{p} \end{pmatrix} \tag{2.6}$$

and the square of the four momentum $\mathbf{p}.\mathbf{p} = \mathbf{p}^2$ is defined as

$$\mathbf{p}.\mathbf{p} = \mathbf{p}^2 = -m^2c^2 + \frac{\lambda^2 l^2}{\hbar^2} p^4 \tag{2.7}$$

using the modified Hamiltonian (2.1). Let us consider the following

$$proton + \gamma = n + \pi^+,$$

that is, a high-energy proton combining with a photon, scatters away the photon from the cosmic background radiation. In the above interaction, the presence of the pion (π^+) allows the conservation of charge. In the lab frame, it is known that the proton requires energy of the order of 10^{20} eV, to ensure that this process occurs. This indeed is the *GZK cutoff*. This limit forbids the detection of cosmic rays with energies $> 10^{20}$ eV as these rays would scatter the CBR photons. However from the modified Hamiltonian (2.1), the energy of the proton E_p is obtained as

$$E_p = 3 \times 10^{20} \text{ eV} + \frac{\lambda_p^2 l_p^2}{\hbar^2} \frac{p^4}{4E_{\gamma}}. \tag{2.8}$$

The first term gives the GZK cutoff and we have $E_p > 10^{20}$. As mentioned earlier, this agrees with the results of AGASA [9] and Hayashida [10]. This maybe seen as the *extended GZK cutoff*. Thus justifying LSV (when the Compton scale is not neglected). Of course, in Eq. (2.8), the conventional GZK effect is obtained if we neglect l_p and set $\lambda_p = 0$. The observed extensions of the GZK effect ensure that the second term in Eq. (2.8) need not be neglected. This is a vindication of the modified dispersion relation (2.1) and can be regarded as a realistic extension of the *classical* special relativity which includes the quantum level.

2.4 Some Effects Due to Neutrino Behaviour

Within the Compton wavelength, the special theory of relativity may not be valid as was shown by the author [2]. Let us consider the example of the neutrino. As known, the mass of the neutrino has not yet been exactly determined, we know only the error in the mass difference squares. The mass of the neutrino, one could say, is roughly 10^{-8} times the mass of the electron. Therefore, the Compton wavelength would be large. This yields a distance of about 10^{-3} cms, corresponding to 10^{-13} s. So, it can be said that for a neutrino, there is an element of uncertainty in position or time of this order. This creates a possibility (in this spacetime interval), for the neutrino to be superluminal. Perhaps, this is a plausible explanation for experiments like MINOS. It is apparent that further observations are called for.

This time shift may be explained with the example of an optical fibre. In an optical fibre, there is a difference in the refractive index (R.I.) of the medium from core to cladding: the refractive index decreases from core to cladding. Near the core, a light ray faces more resistance in the medium, hence causing slower movement, and light rays closer to the cladding move faster because of lesser refractive index. (This in itself could cause some ripple-type interference.)

Now let us consider the case of gamma ray bursts, focussing on an energy band within the burst. After traversing cosmological distances, gamma rays undergo dispersion. This is because of the ZPE, this "viscous" resistance varies between finite values. Thus, there is a frequency variation because there is a slight difference of energies E_1, E_2, etc. This spectra of energies have variations in travel time. There is a time lag due to their frequency distribution.

2.5 Spectral Lags

Let us consider the Snyder–Sidharth energy–momentum dispersion relation. Many authors have proposed a modification of the energy–momentum relation. They have also claimed to account for spectral lags in high-energy gamma rays. A few years ago, Ellis in [15] claimed the existence of events that indicate spectral lags. The author has claimed that Lorentz violation can be observed through a single Compton scattering process, during a simulation of high-energy gamma rays (≥ 1 GeV). In this chapter, the author argues that the GZK limit can be extended beyond 10^{20} eV. It may be reiterated here that so far there is no explanation of the observations of AGASA or Hayashida.

2.6 Spacetime Geometry and Superluminosity

In a recent paper [16], using the noncommutativity of spacetime, the author argued that superluminal velocities were plausible. Superluminosity was also demonstrated experimentally by Megidish et al. [17–19] via the phenomena of entanglement. Let us consider a *gedanken* experiment in which the present and the future maybe viewed as in a periscope.

In this periscope, at the bottom, we have an observer belonging to the present moment and the observer looks through the periscope to find a transparent cube. The periscope is placed along the vertical axis, representing time. The two vertical faces of the cube are considered to be the inner and outer faces. Focussing attention on the outer face it seems that the cube is facing inwards and vice versa (see Fig. 2.1).

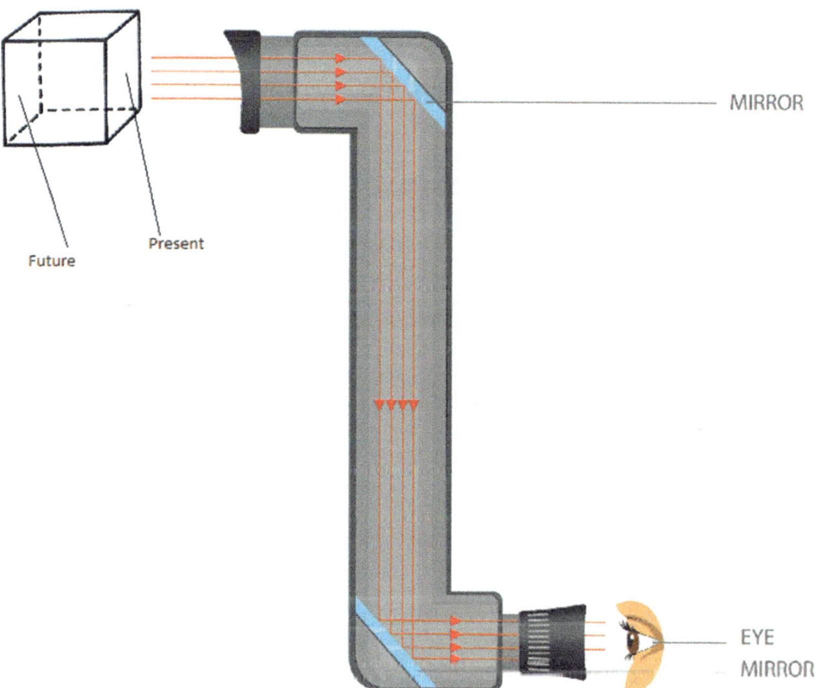

Fig. 2.1 The periscope thought (gedanken) experiment

Perception is created when photons strike the detector. It is as if the photon particles are moving in one temporal direction or the other. This could resemble a Wiener stochastic process. The cube maybe taken to be a trajectory in spacetime between the past and the future. Then, hypothetically the observer perceives from the past into the future, and from the future into the past. For more detailed analysis, see author's work in [20].

Following this line of thought, suppose that an event in the future, as viewed in seen through the periscope is stochastic in the sense of a Wiener process. Let such an event be defined as

$$E_t = \mu t + \sigma W_t,$$

where μ is the drift and σ is the standard deviation. Let us fix these for an event E_t. Then W_t maybe taken to be a Wiener process that constitutes the event. So, transposing the time t is given by

$$t = \frac{1}{\mu}[E_t - \sigma W_t]. \tag{2.9}$$

It may be reiterated that the observer's perception and the faces of the cube are key. Consider the two configurations of the cube, that is, when the cube seems to be projected inward or outward. The parameters μ, σ, and W_t should be such that the value of t corresponds to t', where t' comes from special relativity. In a nutshell, one can say that the parameters' values will be altered.

This basically implies that the event is impacted. But this is meaningless since the event gets altered. In order to overcome this anomaly, that is, to keep the event unchanged, we could consider the time frame to be altered.

This implies that there is a link between the present and future.

In fact, the Wiener processes W_t are involved in the conceptualization of the present–future link. From Eq. (2.9), it is obvious that it is valid for both positive and negative values, that is, t is $\pm ve$. While special relativity forbids violation of causality, the above analysis could still hold good for non-causal events.

Because of the indistinguishability of photons, it makes no difference whether a photon arrives or departs. Basically, this means that an event can occur before it is perceived. That is, when a photon from the observer reaches the event and returns. The positive and negative values of time imply that a stochastic double Wiener process is indicated, that is, a present–future link.

In this scenario, let us invoke a statistical approach. It is known that [21], for many photons with two likely states of "polarizations", the average occupation number is given by

$$n_{\mathbf{k}} = \frac{2}{e^{\beta \hbar \omega} - 1},$$

where $\mathbf{k}$ is the momentum and $\beta = \frac{1}{K_B T}$. Then, the expected number of photons given by the Bose–Einstein statistics is

$$n = \frac{g_i}{e^{(E-\mu)\beta} - 1},$$

where μ is the chemical potential and g_i represents the degeneracy of each energy level i. Let us take the periscope cube to be a scheme wherein the photons are in incessant motion. It is known that photons can vanish into the vacuum when μ (the chemical potential) is set to zero [21]. So, (replacing g_i with g,) if the average occupation number is the same as the expected number of particles, then the two equations above become

$$\frac{g}{e^{\beta E} - 1} = \frac{2}{e^{\beta \hbar \omega} - 1}.$$

Again, since $E = \hbar \omega$, we deduce

$$g = 2 \tag{2.10}$$

this implies that degeneracy is two. This means that there are two states with the same energy level. Given the indistinguishability of photons this could imply the photon has two states.

In one state, the photons reach the cube from the periscope and vice versa in the other state. The photons that originate from the periscope and return from the cube are not distinguishable from the photons that are originating from the cube. Thus making it unclear as to the origin of any particular photon. This is what was termed earlier as present–future correlations.

This gives rise to an interpretation, where the photons themselves are the correlation between the present and the future (with reference to the time axis, along which the periscope is aligned).

2.7 A Second-Order Effect

The author has argued that there could be the so-called Snyder–Sidharth energy–momentum dispersion relation, viz.:

$$E^2 = m^2 + p^2 - 2l^2 p^4. \tag{2.11}$$

In the theory of scattering this further leads to a slight modification of the usual Compton formula, instead of which, we now have

$$k = \frac{mk_0 + \alpha \frac{l^2}{2}[Q^2 + 2mQ]^2}{[m + k_0(1 - cos\Theta)]}, \tag{2.12}$$

where we use natural units $c = \hbar = 1$, m is the mass of the elementary particle causing the scattering, $\vec{k}$, $\vec{k}_0$ are the final and initial momentum vectors, respectively, and $Q = k_0 - k$, and Θ is the angle between the incident and scattered rays. Equation (2.12) shows that $k = k_0 + \epsilon$, where ϵ is a positive quantity less than or equal to $\sim l^2$, l being the fundamental length. It must be remembered that in these units k represents the frequency. The above can be written in more conventional form as

$$\hbar \nu = \hbar \nu_0 [1 + 0(l^2)]. \tag{2.13}$$

Equation (2.13) effectively means that due to the Lorentz symmetry violation in the above theory, the frequency is increased or the speed of propagation of a given frequency is increased. As noted, such models in a purely phenomenological context have been considered by Glashow, Coleman, Carroll, and others [22–24]. In any case what this means in an observational context is that higher frequency gamma rays should reach us earlier than lower frequency ones in the same burst. As Pavlopoulos reports (Cf. Ref. [14]) this indeed seems to be the case [25].

Let us try to find a further test of the extra term in (2.11) by considering Bragg's law which is

$$n\lambda = 2d\sin\Theta. \tag{2.14}$$

It may be remembered that Bragg's law gives the angles for coherent scattering of waves from a large crystal lattice. His law was originally formulated for X-rays, but it also is applicable to all types of matter waves. These include neutron and electron waves provided there are a large number of atoms. In (2.14), d is the distance between atomic layers in a crystal while λ is the wavelength of the incident X-ray beam and finally n is an integer.

We consider the simplest case of a bi-layer crystal of graphene or stanene or similar material. Such materials are not strictly speaking two dimensional but nearly have two layers one above the other. From (2.14), we have

$$n\delta\lambda = 2d\cos\Theta\delta\Theta. \tag{2.15}$$

From considerations leading to Eq. (2.13), we have

$$\delta\lambda \sim L^2, \tag{2.16}$$

where L stands for a typical Compton wavelength, for example, what appeared in (2.11). Using (2.15), (2.16), and (2.14), we get finally

$$\delta\Theta = L^2/d\cos\Theta. \tag{2.17}$$

This shows that there is a deviation $\delta\Theta$ given by (2.17) from the impinging X-rays, due to the second-order effect in Eq. (2.11) or (2.13). It is interesting that

$$d \sim 1\,A^\circ, \tag{2.18}$$

while Θ is an angle which the experimenter can control. In particular, if $\Theta \sim \pi/2$, then we can see from (2.17) that $\delta\Theta$ can take on non-trivial values.

2.8 Back from the Future

Very recently it has been pointed out based on work at several institutes that time entanglement, a purely quantum mechanical effect and also the identity of quantum mechanical particles would lead to a situation where a particle could come back to the present from the future, as described in greater detail below.

Stephen Hawking would time and again say that time travel is possible. Of course all scholars of relativity know how (in principle) one can become younger than his own grandson. Was that what Hawking meant? Recently, however, several *gedanken* performed in Cambridge, Perimeter Institute, Israel, and elsewhere have demonstrated a different effect, based purely on quantum theory. According to this, it is possible to travel into the past and observe the birth of one's own grandfather. These arguments are based on two quantum mechanical inputs. The first is what has recently been termed as time entanglement, rather than spookiness of space entanglement. In other words a future event could in principle influence a past event which seems to go against causality. Another quantum mechanical principle is that all particles of a kind are identical. It is well known that we could substitute one particle by another. Let us see how it works. But first we will touch upon Einstein's spookiness.

2.9 The Spooky Universe

Albert Einstein and Erwin Schrodinger debated a moot point in the 1930s: Einstein's special and general relativity operated in conventional spacetime. Further, superluminal speeds were prohibited by the theories of relativity. What this implies is that there would be causality, that is, for instance, the son could not be born before the father. Surprisingly, some experiments indicated otherwise (for more details about this, see [20]). This made Einstein coin the phrase *spooky action at a distance*. It appeared that the barrier of the velocity of light had been breached. We elaborate this in some little detail. Consider two structureless and spinless particles initially together. These could be particles in a bound state. These would then get separated and move in opposite directions along the same straight line. Measuring the momentum of one of the particles, say, A gives immediately the momentum of the other particle B. The momentum of particle A is equal and opposite to the momentum of particle B. This is because linear momentum is conserved. That this is true in quantum theory too is quite surprising since the momentum of particle B does not have an a priori value, but can only be determined by a separate acausal experiment performed on it.

This can be termed inherent because of the non-locality in quantum theory, arising from the fact that spacetime itself is homogeneous making the conservation of momentum a non-local event. It maybe pointed out that the displacement operator $\frac{d}{dx}$ is, because of the homogeneity of space, independent of x. This results in the conservation of momentum in quantum theory (cf. Ref. [26]). The displacement δx which leads to $\frac{d}{dx}$ represents an instantaneous shift of origin corresponding to an infinite velocity and is compatible with a closed system.

It can also be true if the instantaneous displacement is taken to be a real displacement in actual time δt. This is a possibility when the states are stationary, which is the case when the total energy is constant.

Here the space and time displacement operators, which usually are not on the same footing, in this case maybe treated on par [27]. It maybe observed that in relativistic quantum mechanics, x and t are on an even keel. Also to be borne in mind is the fact that special relativity considers inertial or unaccelerated frames.

In general, in field theory, we consider different space points but at the same instant of time. However, because of the finite velocity of light, this information is retrieved for different instants of time. If information is needed for the same instant of time, it is possible only for stationary states. Let us consider now the field equations which are derived from a variational principle:

$$\delta I = 0. \tag{2.19}$$

In this deduction, the δ operator which refers to an arbitrary variation commutes with the space and time derivatives. That is, the momentum and energy operators constitute a complete set of observables. As such the apparently arbitrary operator δ in (2.19) is constrained to be a function of these (stationary) variables [28].

All this highlights the following facts: first we are implicitly dealing with an a priori homogeneous space, that is, the physical space. Secondly, in the relativistic picture the space and time coordinates are taken to be on the same footing, which they are not, as noted by Wheeler [29]. Our concept of the universe is based on "all space (or as much of it as possible) at one instant of time".

It must be mentioned, though, that in conventional theory this is only an approximation. In the author's conceptualization, the particles are fluctuationally created out of a background ZPE. It is these N particles that define physical space, which makes it no longer a priori as in the Newtonian formulation. It is only in the thermodynamic limit in which, when $N \to \infty$ and $l \to 0$, that we recover the above classical picture of a rigid homogenous space, with the conservation laws.

These conservation laws, strictly speaking, hold in what is called the thermodynamic limit, which means that the number of particles tends to infinity. That is, effectively, N is taken to be very large.

This gives a description of a cosmology where $\sqrt{N}$ particles are created by fluctuations in the background ZPE. This implies, because of the fluctuations, that the violation of energy conservation is proportional to $\frac{1}{\sqrt{N}}$. Following a similar line of thinking, the violation of momentum conservation may be taken to be proportional to $\frac{1}{\sqrt{N}}$ (per particle).

Therefore, there is a very small probability that the measurement on particle A will not lead to exact information about particle B.

From classical concepts of spacetime, this includes the theories of relativity this would be relevant. Since from classical theory it is possible to have information about two particles separated in space at the same instant of time. This is not the case in quantum theory, because we can ascertain a particle's position and velocity only by making a measurement for one particle and a separate measurement for the second particle. Hence, Einstein characterized this "as spooky action at a distance". Schrodinger, on the other hand, had a different take on this issue. He introduced the concept of non-separability. According to him, if two particles interact at an instant of time, they continue to remain "entangled" for all time. This means that, in principle, it is possible to obtain measurements of the momentum and position of the second particle also without any contradiction. What is being said is that there is no clash with the so-called concept of "spookiness". Or the two particles are no longer independent, they remain "entangled". This is an over-simplification of space entanglement.

On the other hand, time entanglement requires a lot more subtlety as is evidenced from past discussions by the author and others. Many *gedanken* have been proposed by scholars: the Hebrew University of Jerusalem, the University of Cambridge, and so on. All these thought experiments relate to this phenomenon of time entanglement. Specializing to the time event, intuitively, an event taking place now can exert influence on an event taking place later on. Strangely enough work suggests the opposite: that it is the future which influences the past. In other words, the particle in the future may influence the particle in the present. It was reported by Megidish and his colleagues that in some bizarre manner the future influences the present or past! Or they contended that there was entanglement between photons separated in time. These physicists at the Hebrew University of Jerusalem reported in 2013 that they had successfully performed an entanglement experiment with photons that did not coexist at all. Earlier experiments used a method called "entanglement swapping". This had demonstrated quantum correlations across time. They delayed the measurement of one of the coexisting entangled particles. Eli Megidish and his collaborators were the first to show entanglement between photons which never coexisted. The entanglement protocol entangles two distant non-interacting photons.

They used the following technique:

First, they entangled a pair of photons, say, 1 and 2. Then they measured the polarization of photon 1 (a property describing the direction of light's oscillation) thus sending it to an eigenstate (step II).

Photon 2 was sent on an irrelevant trajectory, and, at the same time, a new entangled pair, 3 and 4, was created (step III).

A measurement was then made on photon 3 and photon 2 in such a way that the entanglement relation was swapped from the old pairs (1 2 and 3–4) onto the new 2–3 combo (step IV).

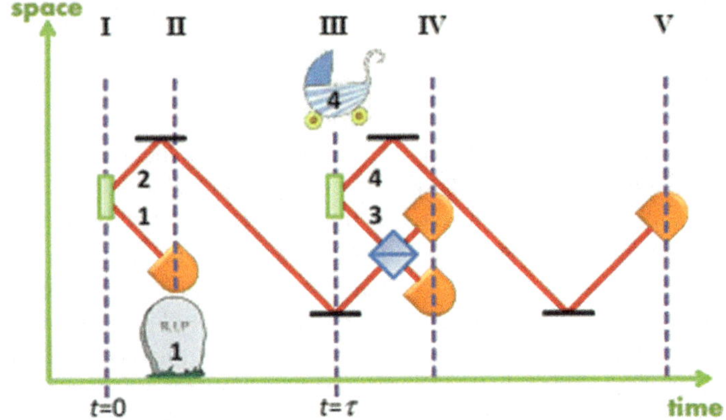

Fig. 2.2 Time line diagram. (I) birth of photons 1 and 2. (II) detection of photon 1. (III) birth of photons 3 and 4. (IV) Bell projection of photons 2 and 3. (V) detection of photon 4 (picture courtesy [17])

Sometime later (step V), the polarization of the surviving photon 4 was measured, and the results were compared with those of the long-gone photon 1 (back at step II). To sum up, the existence of quantum correlations between temporally non-local photons 1 and 4 was shown (see Fig. 2.2) which means that entanglement can take place for two quantum systems which, in the first place, had no coexistence.

Finally, we would like to make the following comments[1]:

1. In the above scheme of things, causality exists although in a modified sense.
   An event could be caused by another event or could also cause it. This phenomenon can be pictured by a new Feynman-type "spacetime diagram". Let us consider a rectangular matchbox configuration for example (this has been discussed in (Cf. Ref. [20])). The two faces of this matchbox are, say, ABCDE and FGHJI. These are either facing us or facing away. Both these are true depending on which face we see first. In the present instance, seeing first relates to making an observation. So both possibilities are equally probable.

2. The author, as noted several times, proposed in 1997 a dark-energy-driven universe that would be accelerating [25]. The characterization here is of a universe embedded in a swathe of dark energy. This is somewhat reminiscent of the model of Nernst in the early 1900s. This dark energy model of the universe would dispense with both the Liebnitzian and Newtonian concept of spacetime. This leads to a satisfactory holistic and Machine view of the universe which has no conflict with new phenomena.

3. It must be observed that in the author's work [25] the usual commutators of quantum physics are replaced by the so-called Snyder–Sidharth relations. With these any two quantum mechanical observables like space coordinates, momentum coordinates, or spacetime have non-zero commutators even though the effect

---

[1] Talk delivered at the University of Udine, May 2019.

maybe miniscule. This view has shades of Dirac's characterization of commutativity and compatibility [26]: where two commuting observables could be simultaneously measured as long as they were not non-commutating. So, with this new degeneracy, entanglement would spread all across the universe, macro as well as micro.

4. Finally, it may be observed that the author has a completely different characterization of spacetime [30]. In this characterization, the author brings out a correspondence between spacetime intervals and the displacements of a random walk. In a random walk, as is known, even when the number of forward steps is equal to the backward steps there is still a non-zero displacement. And it is this displacement which defines space and time intervals.

All the above considerations make time entanglement appear less incredible than what it might seem at first sight.

References

1. B.G. Sidharth, Discrete spacetime and Lorentz symmetry. Int. J. Theor. Phys. **43**(9), 1857–1861 (2004)
2. B.G. Sidharth, Different routes to Lorentz symmetry violations. Found. Phys. **38**(1), 89–95 (2008)
3. S. Sahoo, Mass of the neutrino. Indian J. Pure Appl. Phys. **48**, 691–696 (2010)
4. R. Andriambololona, C. Rakotonirina, A study of the Dirac-Sidharth equation. EJTP **8**(25), 177–182 (2011)
5. B.G. Sidharth, A Model for Neutrinos. Int. J. Theor. Phys. **52**, 4412–4415 (2013)
6. J. Albert, E. Aliu et al., Variable very high energy γ-ray emission from Markarian 501. Astrophys. J. **669**(2), 862 (2007)
7. W. Bednarek, R.M. Wagner, A model for delayed emission in a very-high energy gamma-ray flare in Markarian 501. Astron. Astrophys. **486**(3), 679–682 (2008)
8. T.N. Ukwatta, M. Stamatikos et al., Spectral lags and the lag-luminosity relation: an investigation with swift bat gamma-ray bursts. Astrophys. J. **711**(2), 1073 (2010)
9. K. Shinozaki, M. Teshima, Agasa Collaboration, et al., AGASA results. Nucl. Phys. B-Proc. Suppl. **136**, 18–27 (2004)
10. N. Hayashida, K. Honda, et al. Observation of a very energetic cosmic ray well beyond the predicted 2.7 k cutoff in the primary energy spectrum. Phys. Rev. Lett. **73**(26), 3491 (1994)
11. B.G. Sidharth et al., The anomalous gyromagnetic ratio. Int. J. Theor. Phys. **55**(2), 801–808 (2016)
12. B.G. Sidharth et al., Revisiting the Lamb shift. Electron. J. Theor. Phys. **12**(34), 15–34 (2015)
13. B.G. Sidharth, *The Universe of Fluctuations* (Springer, 2005)
14. T.G. Pavlopoulos, Are we observing Lorentz violation in gamma ray bursts? Phys. Lett. B **625**(1–2), 13–18 (2005)
15. J. Albert, E. Aliu et al., Probing quantum gravity using photons from a flare of the active galactic nucleus Markarian 501 observed by the MAGIC telescope. Phys. Lett. B **668**(4), 253–257 (2008)
16. B.G. Sidharth et al., Spacetime geometry and the velocity of light. Int. J. Mod. Phys. A **33**(30), 1850183 (2018)
17. E. Megidish, A. Halevy, T. Shacham, T. Dvir, L. Dovrat, H.S. Eisenberg, Entanglement Swapping between Photons that have never coexisted. Phys. Rev. Lett. **110**, 210403 (2013), (May)

18. X.S. Ma et al., Experimental delayed-choice entanglement swapping. Nat. Phys. **8**(6), 479–484 (2012)
19. E. Megidish, T. Shacham, A. Halevy, L. Dovrat, H.S. Eisenberg, Resource efficient source of Multiphoton Polarization Entanglement. Phys. Rev. Lett. **109**, 080504 (2012). (Aug)
20. B.G. Sidharth, *Chaotic Universe* (Nova Science, New York, 2001)
21. K. Huang, *Statistical Mechanics* (Wiley Eastern, New Delhi, 1975)
22. S.M. Carroll, G.B/ Field, R. Jackiw. Limits on a lorentz-and parity-violating modification of electrodynamics. Phys. Rev. D **41**(4), 1231 (1990)
23. S. Coleman, S.L. Glashow, High-energy tests of lorentz invariance. Phys. Rev. D **59**(11), 116008 (1999)
24. S.L. Glashow, Breaking lorentz invariance, in *AIP Conference Proceedings*, vol. 444.(American Institute of Physics, 1998), pp. 119–129
25. B.G. Sidharth, *The Thermodynamic Universe* (World Scientific, Singapore, 2008)
26. P.A.M. Dirac, *The Principles of Quantum Mechanics*, vol. 27 (Oxford university press, 1981)
27. A.S. Davydov, *Quantum Mechanics* (Elsevier, 1966)
28. B.G. Sidharth, Quantum nonlinearities in the context of black holes. Nonlinear World **4**, 157–162 (1997)
29. C.W. Misner, K.S. Thorne, *and J* (Freeman, A. Wheeler. Gravitation, 2000)
30. B.G. Sidharth, spacetime as a random heap. Chaos, Solitons & Fractals **12**(1), 173–178 (2001)

Chapter 3
The Enigmatic Neutrino

3.1 Introduction

Ever since its postulation by Wolfgang Pauli in the 1930s and subsequent discovery, the neutrino has been an enigmatic particle that has defied our complete understanding. The most prominent feature was its handedness, which made it a unique particle and a vital player in weak interactions. When the standard model of particle physics was completed around 1970, the neutrino was considered to be a massless particle. In the 90s, the author suggested a small mass for the neutrino [1]. Later alternative mechanisms for the neutrino mass were postulated [2]. This was confirmed in 1999–2000 by the Super-Kamiokande experiment in Japan.

This made the neutrino (with all its flavours) an even more problematic particle. It had a mass, miniscule albeit, but also it travelled with the speed of light. Nevertheless neutrino physics and astrophysics are fields of continuing intense study. We shall see some aspects of this below.

3.2 A Background Neutrino Model for the Cosmological Constant

As mentioned earlier, in 1997, the author came up with a model of a universe which was driven by dark energy. In this theory, the cosmological constant had a much smaller value than was the accepted one. The generally accepted view of universe at that period was in fact the opposite [1]. The author's concept of an expanding universe was, in 1998, confirmed by the observations of Perlmutter and others, who came to this conclusion after observing distant type Ia supernovae.

As noted in Chap. 1, the characterization of dark energy is a mystery to all scientists, although the concept of dark energy is widely accepted. There are different approaches to the concept of dark energy by different scholars.

B. G. Sidharth, *The Dark Energy Paradigm*,
https://doi.org/10.1007/978-981-96-3745-4_3

- The cosmological constant concept.
- Another is to identify dark energy with a scalar field, for instance, quintessence. Quintessence is considered to be a hypothetical form of dark energy, a scalar field, proposed to provide an explanation of the accelerating rate of a universe which was expanding. Such a field can also be associated with a fundamental particle or a composite particle.
- Tachyonic fields are another candidate [3–5].
- As an example, an interaction of dark energy with a fermionic field, contained in dark matter, could be considered, these fermions being neutrinos [6, 7].
- Some authors [8] have attempted to formulate an equation of state for a dark energy fluid.
- There have also been considerations about whether cosmology is dissipative or otherwise [9].
- Some scholars utilize a generalized second law for this purpose [10]. We had touched upon this earlier.
- Another angle is via the coincidence problem, viz. why the energy density of dark energy is roughly of the same order as a cosmological critical density [11–13].

In any case, as we renormalize, effectively an infinite amount of energy is neglected. At this juncture one could question whether dark matter exists or not. As is well known dark matter has not been observed after its existence was postulated decades ago. What dark matter is, is still unknown. What can be said is that there are alternative explanations which do not need dark matter. The author, in his cosmology, uses the gravitational constant that depends inversely on time. In this theory, the time-varying gravitational constant can explain the anomalous phenomena attributed to dark matter. One such is the galactic rotation curves anomaly (Cf. Ref. [14] and several references therein). With the use of a time-varying gravitational constant a uniform cosmic acceleration is explained with an approximate order of $10^{-7}\,\mathrm{cm/s^2}$. One could say that this approach could replace the modified Newtonian gravity approach (MOND).

In the context of cosmic background neutrinos [15] let us examine dark energy. The author had shown earlier [16–18] within the framework of the cold cosmic neutrino background that the neutrino mass and other neutrino parameters could be determined. The values of the neutrino mass thus computed are within the limits obtained from the Super-Kamiokande experiments. This work predicted the results [19] of the Super-Kamiokande experiments. To see this, consider the cold Fermi degenerate gas. In this case, we have the following:

$$p_F^3 = \hbar^3 (N/V), \tag{3.1}$$

where p_F is the Fermi momentum, N is the number of particles, and V is the volume of the gas.

To deduce (3.1) (Cf. Ref. [20]), we use the fact that in the ground state of a cold degenerate Fermi gas, the particles occupy the lowest possible energy levels and fill all levels up to the Fermi energy e_F. Alternatively, in momentum space, the particles fill a sphere of radius p_F. So, the number N of these particles is given by

$$\frac{V}{\hbar^3} \int_{e_p < e_F} d^3 p = N. \tag{3.2}$$

As

$$e_F \sim p_F^2 / m$$

we find that p_F must satisfy Eq. (3.1).

Substituting the neutrino parameters which are known, viz. [21] $N \sim 10^{90}$ we get consistently the neutrino mass $\sim 10^{-3}$ eV [22]. The background temperature $T \sim 1°K$ as KT is the Fermi energy e_F. Recently, physicists have been looking for the ripples of the early big bang in this neutrino background, as has been claimed in the work of Trotta and Melchiorri [23].

There is growing evidence for cosmic background neutrinos (see, for example, the work of Weiler [24]). It can be safely conjectured that the GZK photo pion process is the contributing factor.

With this scenario we derive the cosmological constant from the Fermi energy of the cold neutrino background. We begin with the following:

$$\text{Fermi Energy} = \frac{N^{5/3}\hbar^2}{m_\nu R^2} = M \Lambda R^2, \tag{3.3}$$

where M is the mass of the universe, R is radius $\sim 10^{27}$ cm, and Λ is the cosmological constant m_ν, the mass of the neutrino.

Several years ago Hayakawa [17, 25] used the expression $\frac{N^{5/3}\hbar^2}{m_\nu R^2}$ in (3.3) for the cold background neutrino Fermi energy. However, he did not use the cosmological constant. He had a suitable counter-balancing gravitational force.

From (3.3) we obtain a value for the cosmological constant,

$$\Lambda \sim 10^{-37} \, \text{s}^{-2} \tag{3.4}$$

which is of the correct order [26]. The author has derived the same result from another perspective [27].

The cosmological constant is given by

$$\Lambda = \langle 0 | H | 0 \rangle \equiv \text{cosmological constant}. \tag{3.5}$$

From which the cosmological constant Λ is obtained by its familiar expression [28] $(c = 1)$

$$\Lambda = \int_0^L \frac{4\pi p^2}{(2\pi)^3} dp \frac{1}{2}\sqrt{p^2 + m^2}. \tag{3.6}$$

In (3.6) $L \sim p_F$ is the cutoff which makes it possible to take care of the divergent integral. Using the value of the neutrino mass $\sim 10^{-3}$ eV in (3.6) we obtain the cosmological constant as

$$\Lambda \sim 10^{-48}\,\text{GeV}^4 \tag{3.7}$$

which is the same as (3.4).

Whereas conventional quantum field theory or string theory throws up a cosmological constant which is over a hundred orders of magnitude than this observed value, as already noted.

3.3   Neutrino Mass

In the year 2000, the author had proposed a modified form of special relativity, which takes into account the fact that there are no spacetime points but rather only discrete spacetime intervals which are nevertheless miniscule. This leads to a modified energy–momentum formulation, the so-called Snyder–Sidharth Hamiltonian:

$$E^2 = p^2 c^2 + m^2 c^4 + \frac{\alpha c^2}{\hbar^2} l^2 p^4, \tag{3.8}$$

where l is a minimum length like the Planck length, p is the magnitude of the momentum vector $\vec{p}$, and α is positive for fermions or spin half particles like neutrinos [1, 14, 29–31]. This equation treats spacetime as nondifferentiable at ultra-high energies. This, as mentioned earlier, is a deviation from conventional quantum theory with spacetime as points. In the modified scenario, the commutativity in quantum theory is replaced. This was pointed out by Snyder several decades ago (Cf. Refs.[1, 30]) by

$$[x, y] = O(l^2), [x, p_x] = \iota \hbar [1 + l^2]. \tag{3.9}$$

Equation (3.8) shows that at very high energies, the energy of fermions is greater than that which is thrown up by the special theory of relativity. This greater energy implies a superluminosity for fermions or the extra term in (3.8) can be treated as an extra mass or possibly energy that can be identified with Cherenkov radiation. It may be pertinent to mention that [32] Cherenkov surface waves consist of photonic quasi-particles and propagate in two dimensions. These are emitted by free electrons. The reduced dimensionality is predicted to change the properties of free electron radiation. The results of the work in [32] concur with the theoretical prediction of the author that free electrons do not always emit classical light. These maybe entangled with the photons which they emit.

This, in fact, is the mass-generating mechanism, as pointed out by the author [33, 34].

That is, the neutrino, which is supposedly massless, develops mass, as can be seen from Eq. (3.8). This mass turns out to be of the correct order.

3.4 Neutrino Mass Measurements

- The Standard Model predicts that neutrino masses are exactly zero. However, the discovery of neutrino oscillations disproved the Standard Model as far as neutrino masses were concerned. Neutrino oscillation measurements did not facilitate obtaining the scale of neutrino masses. Neutrino mass scale is obtained from the energy spectrum created by beta decay. Several measurements of neutrino mass have been made. See for example the review paper for the direct measurements of neutrino mass, *Direct measurements of neutrino mass* [35]. It has been seen from the measurements of the flux of solar neutrinos that there is a deficit of electron-type neutrinos which reach the Earth. This was the so-called solar neutrino problem. The data obtained shows that electron neutrinos are linear superpositions of at least two neutrino mass eigenstates. With one of these states having non-zero mass, the difference between the neutrino masses-squared is obtained to be of the order $\Delta_{21}^2 \equiv m_2^2 - m_1^2 \sim 10^{-4} eV^2$. Here m_1 and m_2 are the masses of two different neutrino mass eigenstates.
- For another perspective of understanding of neutrino masses and mixings, see [36]. The authors, in this review paper, survey the present and future outlook of the understanding of neutrino masses and mixings. They attempt to explain the flavour problem of quarks and leptons.

3.5 Some Salient Aspects of Neutrinos

1. Cherenkov radiation occurs when electrically charged particles travel faster than light in a medium, causing the release of photons. Therefore, Cherenkov radiation should not apply to neutrinos because they are electrically neutral. The conventional view has been that neutrinos were electrically neutral fermions that interact with gravity and the weak interaction. They have a very small rest mass and typically pass through normal matter undetected. On the other hand, the author has proposed, based on his theory of Fluctuational Cosmology [16], that the neutrino can have an effective small "charge", $\bar{g}$, given by

$$\bar{g}^2/e^2 \sim 10^{-13}. \tag{3.10}$$

This would mean that neutrinos would have a faint Cherenkov effect.

2. Bosons can be bound states of fermions, as proposed by the author and several others [1] and also Klauder [37] and others such as [38]. Consider a composite particle consisting of simple particles bound together by a potential. This can be a boson or a fermion, depending on the distance between the particles. The particle behaves like its constituent particles at small distances. At large distances, the particle behaves as a boson or fermion. Fermions can exhibit bosonic behaviour

when they are bound loosely in pairs. For instance, electrons in superconducting materials form Cooper pairs through the exchange of phonons.
Consider the wave function given by [39], with respect to neutrinos

$$\chi_j = E_j + \iota B_j, \ \chi_0 = 0, \tag{3.11}$$

where χ_j is the wave function and E_j represents the electric field, and B_j the magnetic field. The Maxwell equations then become

$$\beta_\mu \frac{\partial \chi_\nu}{\partial x_\mu} = -\frac{1}{c} j_\nu \tag{3.12}$$

resembling the Dirac equation. Thus, the Cini–Toushek–Dirac equation which is the zero mass Dirac equation gives back Maxwell's equations. The photon could be viewed as a combination of a neutrino and anti-neutrino. This also resembles an ultra-high-energy electron. Both of these can be represented as two-component spinors. To represent this we use the following combination

$$D^{(\frac{1}{2}0)} \oplus D^{(0\frac{1}{2})} \tag{3.13}$$

rather than the regular bound state.

3. The author proposed in [15, 40] that the neutrino is a two-dimensional object. This is apparent from the two-component neutrino equation itself. Interestingly, this very equation maybe used to describe particles or quasi-particles in graphene, a two-dimensional form of graphite. This maybe seen as follows: From the Dirac equation, considering only positive-energy solutions or negative-energy solutions alone [41], we get

$$\psi^{(+)}(x, t) = \int \frac{d^3 p}{(2\pi\hbar)^{1/2}} \sqrt{\frac{mc^2}{E}} \sum_{\pm s} b(p, s) u(p, s) e^{-\iota p_\mu x_\mu / \hbar}. \tag{3.14}$$

The expectation of the velocity operator is given by

$$\mathbf{J}^{(+)} = \int \psi^{(+)\dagger} c\vec{\alpha}\psi^{(+)} \mathbf{d}^3\mathbf{x} \tag{3.15}$$

which yields

$$\mathbf{J}^{(+)} = \langle c\alpha \rangle_+ = \left\langle \frac{\mathbf{c}^2\mathbf{p}}{\mathbf{E}} \right\rangle_+ = \langle \mathbf{V}_{\mathbf{gp}} \rangle_+, \tag{3.16}$$

where $\langle \rangle_+$ denotes the expectation with respect to a positive-energy packet. Clearly (3.16) shows that the velocity equals the speed of light. Also since

$$[H, c\vec{\alpha}] \neq 0,$$

the velocity is variable. It is interesting to point out here that Dirac noted this in the process of deriving the relativistic electron equation [42]. His rationale was that all measurements are averages for a short interval of time. This short interval would be of the order of the Compton time. In this short span, the *zitterbewegung* effects would be averaged out and a sub-luminal value for the velocity would emerge. Equivalently this is to say that a wave packet consisting of solutions of one sign moves with velocities equal to that of light. Thus a wave packet with both positive and negative energies would be needed to obtain the usual sub-luminal velocities and mass (for more details see [43]).

So, what we are saying here is that the neutrino can be described as consisting of solutions with one sign of energy. This is the reason for its luminal speeds and near-vanishing mass. This implies that the Dirac spinor equation which consists of four components becomes a two-component one. Each of these components consists of two positive solutions and two negative solutions. That is, the Dirac spinor becomes a positive-only or negative-only chiral spinor. This of course is the familiar Weyl equation which represents neutrinos as two-component objects. Several features including the near masslessness chirality are easily seen. Also, the two-dimensional feature of neutrinos could explain their oscillations.

4. The author, [44], a few years ago made a case that the neutrino and anti-neutrino are asymmetric. The same concept has also been worked out by others. For example, see [45]. Of course this would lead a charge-parity violation. Experiments at the MINOS of Fermilab and the T2K experiment (Tokai to Kamioka) corroborate this [46, 47]. The MINOS or Main Injector Neutrino Oscillation Search experiment is a neutrino experiment designed to observe neutrino oscillations. Firstly, the rate at which muon neutrinos disappear into other types is measured. Second, they measure whether those other types of neutrinos are known. Third (the most relevant to this work), they search for the appearance of electron neutrinos from muon neutrinos. This oscillation is perhaps the best way to measure CP-violations. This could also be at the root of the matter–antimatter asymmetry in the universe. As is known, this would imply that a mass is required for the oscillations.

5. The energy–mass relation of special relativity maybe confirmed within an error $\sim 10^{-7}$ [48].

   The author in Ref. [48] tests the mass–energy relationship directly by combining very accurate measurements of atomic-mass difference, Δm and of γ ray wavelengths to determine E, the nuclear binding energy, for isotopes of silicon and sulphur. The author of this study found that the energy mass formula can be separately confirmed in two tests yielding a combined result of $1 - \Delta mc^2/E = (-1.4 \pm 1.4) \times 10^{-7}$, indicating that it holds to a level of at least 0.00004%.

   Using this in the modified dispersion relation (3.8) within the limit of this error we have that the last term in (3.8) becomes of the order less than or $\sim 10^{-7}$. This aids in the calculation of the mass to be $\sim 10^{-8}$ times the electron mass. This interestingly is the order the neutrino mass itself! In fact it is the order of the mass difference squared. This could be an alternative derivation. We return to this point later.

3.6 A Model for Neutrinos

3.6.1 Mass, Charge, Chirality, and Other Considerations

In 1930, Wolfgang Pauli proposed the existence of a new tiny particle with no electric charge. The particle was postulated to be very light or maybe massless. It took more than 25 years for scientists to discover that neutrinos exist though they are everywhere. Since then, it has been found that its mass is $\sim 10^{-8}$ of the electron mass. Actually what has been observed is the mass differences between different flavours of neutrinos. Neutrinos turn out to be fermions travelling with almost the speed of light. It is estimated that there are some 10^{90} neutrinos in the universe. Recently, a cosmic neutrino background has also been detected [24]. The two-component Weyl equation aptly describes the neutrino unlike other fermions. The neutrino also exhibits chirality. According to John Wheeler, it is mandatory for the neutrino to have a general relativistic description. Unless this is so, its unification with quantum theory would not be possible [49]. From July 2006 to December 2012, the *CERN Neutrinos to Gran Sasso* (CNGS) project sent muon (μ) neutrinos from CERN to the Gran Sasso National Laboratory (LNGS), 732 km away in Italy. The muon neutrinos changed to tau (τ) neutrinos in the CNGS beam as observed by the OPERA experiment. That is, μ neutrinos had switched flavour to appear as τ neutrinos at Gran Sasso. The question that comes up is whether such effects could solve the well-known problem of the solar neutrino deficit.

3.6.2 Two Dimensionality

A slightly heretic path is now boldly proposed. The author has been arguing for several years that the neutrinos are two-dimensional entities in some sense. This would possibly explain their maverick-like features. Firstly, the Dirac equation in two space dimensions maybe used to represent the neutrino. As was pointed out by the author a couple of decades ago [50, 51]. Furthermore, recently, this again resurfaced because of the work of the author on graphene, which is a two-dimensional sheet of graphite loosely. In this context, quasi-particles with distinctly neutrino-like properties are encountered [52–54].

To see this let us linearize the relativistic energy–momentum relation:

$$E^2 = p^2 c^2 + m_0^2 c. \tag{3.17}$$

Here, $\vec{p} \cdot \vec{p} = p^2$, p representing the magnitude of $\vec{p}$, and then the corresponding quantum mechanical equation is

$$H\psi = [c\vec{\alpha} \cdot \vec{p} + \beta m_0 c^2]\psi \tag{3.18}$$

where, as seen

$$\vec{p} \equiv \frac{\hbar}{\imath}\left(\frac{\partial}{\partial x}\,\frac{\partial}{\partial y}\right) \text{ and } H \equiv \frac{\hbar}{\imath}\frac{\partial}{\partial t}.$$

Multiplying (3.18) by H on the left side and $(c\vec{\alpha}\cdot\vec{p}+\beta m_0 c^2)$ on the right side and comparing with Eq. (3.17), we get

$$(\alpha^\imath\alpha^j + \alpha^j\alpha^\imath) = 2\delta^{\imath j},\, \alpha^\imath\beta + \beta\alpha^\imath = 0,\, \beta^2 = 1,\, \imath j = 1, 2. \tag{3.19}$$

Equation (3.19) is satisfied if the set $(\vec{\alpha}, \beta)$ is the set of Pauli matrices.

Thus, Eq. (3.18) is clearly a two-component fermion whose relativistic covariance can be easily established. It has been the view of some authors that the neutrino is a two-component fermion [55]. Furthermore, the reason that the neutrino satisfies the Dirac equation or the Weyl equation (which is a special case of the Dirac equation having two components) is that the rest mass m_0 of the neutrino vanishes.

So if a fermion can be confined to two dimensions, say, the surface of a thin strip of negligible thickness, possibly graphene, then it should behave like the massless and parity-violating neutrino.

It maybe pointed out that the cosmic background neutrinos are ultra-cold being at temperature $T \approx 3°K$ which makes the collection of cosmic background neutrinos nearly mono-energetic. This leads to a bosonic behaviour. The argument is as follows. We start with the familiar expression for the occupation number of a fermion gas [20]

$$\bar{n}_p = \frac{1}{z^{-1}e^{bE_p} + 1}, \tag{3.20}$$

where $z' \equiv \frac{\lambda^3}{v} \equiv \mu z \approx z$ because, here, as can be easily shown $\mu \approx 1$,

$$v = \frac{V}{N},\, \lambda = \sqrt{\frac{2\pi\hbar^2}{m/b}}$$

$$b \equiv \left(\frac{1}{KT}\right), \quad \text{and} \quad \sum \bar{n}_p = N. \tag{3.21}$$

Let us consider a hypothetical, nearly mono-energetic collection of fermions. Its distribution is given by

$$n'_p = \delta(p - p_0)\bar{n}_p, \tag{3.22}$$

where $\bar{n}_p$ is given by (3.20).

Let us consider a special collection of nearly mono-energetic particles in equilibrium so that we simulate the cosmic background neutrinos.

By the usual formulation, we have

$$N = \frac{V}{\hbar^3} \int d\vec{p}\, n'_p = \frac{V}{\hbar^3} \int \delta(p - p_0) 4\pi p^2 \bar{n}_p dp = \frac{4\pi V}{\hbar^3} p_0^2 \frac{1}{z^{-1} e^{\theta} + 1}, \quad (3.23)$$

where $\theta \equiv b E_{p_0}$.

The δ function in (3.22) in momentum space causes a reduction in dimension. This could be considered a fractal two dimension. In the relativistic case, this could lead to anomalous behaviour (see [56] for details). Another perspective would be that dimensionality itself is connected to the virial distribution of velocities. In any case, in this situation, there is no such a velocity spread. In recent quantum gravity approaches [57], this two-dimensional aspect has been considered via the holographic principle. That is, three dimensionality is an artefact of observation as in a hologram. All aspects of the image are contained in two dimensions. Another example is a black hole, where all information is confined to the surface. The author has been postulating that the universe itself can be characterized as a black hole [14]. Also, in 1996, the author (and A.D. Popova) had hypothesized that the universe is asymptotically two dimensional.

An alternative perspective would be to observe the dynamics of the rotation curves of galaxies [58]. Observations show that for galaxies with a large radius R, the mass M has the relation:

$$M \propto R^n, \quad \text{where } n \approx 2$$

approximating two dimensionality.

All this bolsters the case for the neutrino to be treated as a two-dimensional object which satisfies the two-dimensional Dirac or Weyl equation [40].

Succinctly, a wave packet of solutions with one sign of energy moves with the velocity of light. A wave packet with both positive and negative energies would be needed (as mentioned earlier) to have the usual sub-luminal velocities and mass. This maybe seen in [43].

Thus to reiterate, the neutrino can be described as consisting of one sign of energy solutions only.

This is the reason for its speed of light and mass which is nearly vanishing. This also means, as noted, that the Dirac spinor equation consisting of four components, two of which are positive solutions and two negative solutions, is a two-component wave function, with two spinors, that is, a positive-only or a negative-only spinor. We are alluding to the Weyl equation which represents neutrinos as two-component objects as seen earlier. Several features like the near masslessness and chirality are explained.

We know that a particle localized in space can be represented by a wave packet containing both positive-energy and negative-energy solutions. Those particles with energy solutions of a single sign, positive or negative, would travel with the velocity of light and would hence be massless. Interestingly, this would mean that the Compton length of such particles, here we are talking about neutrinos, would be infinite or practically speaking very very large [41, 43, 59, 60].

The author derives the cosmological constant from cosmic neutrino background [61]. Beginning with the cold Fermi degenerate gas we have (as mentioned in Sect. 3.2)

$$p_F^3 = \hbar^3 (N/V). \tag{3.24}$$

This can be deduced by using a property of the ground state of such a Fermi assembly. In this the neutrinos occupy the lowest possible energy levels. Of course all energy levels up to the Fermi energy ε or e_F are occupied. This means that the neutrinos fill a sphere of radius p_F. So we have

$$\frac{V}{\hbar^3} \int_{e_p < e_F} d^3 p = N \tag{3.25}$$

using

$$e_F \sim p_F^2/m$$

(3.24) follows.

Invoking the neutrino parameter, viz. [21] $N \sim 10^{90}$ we get a consistent neutrino mass $\sim 10^{-3} eV$ [22]. Here the background temperature $T \sim 1^\circ K$ as KT is the Fermi energy e_F.

These days it is hoped that neutrinos also display the ripples of the early big bang, a result claimed by Trotta and Melchiorri [23].

The GZK photo pion process seems to yield evidence for cosmic background neutrinos [24].

We now recall the expression for the Fermi energy

$$\text{Fermi Energy} = \frac{N^{5/3}\hbar^2}{m_\nu R^2} = M \Lambda R^2, \tag{3.26}$$

where M is the mass of the universe, R is radius $\sim 10^{27}$ cm, and Λ is the cosmological constant and get from (3.26):

$$\Lambda \sim 10^{-37}\, s^{-2} \tag{3.27}$$

(3.27) which gives the cosmological constant of the right order.

It maybe pointed out at this juncture that using conventional arguments a hugely wrong value of the cosmological constant which is 10^{120} times the order of magnitude of the observed cosmological constant is obtained.

This suggests that Einstein was to a certain extent vindicated: to the extent that neutrinos are two-dimensional objects and can be treated as "part" of radiation.

3.6.3   Two-Dimensional Neutrinos

We next come to Dirac's characterization of quantum mechanics. This arises if there is a minimum space or time interval which cannot be penetrated. Then we transit from classical physics to quantum physics. In other words, infinite precision implies classical physics. Dirac in [42] elaborated as follows: using *zitterbewegung* which is rapid oscillation that takes place within the Compton wavelength or Compton time and once averages over this interval are taken, the *zitterbewegung* disappears and we get back classical physics with an appendix which is the *zitterbewegung* region that is imaginary. Remembering that the Compton wavelength is given by $\hbar/mc$. This shows that a mass is thrown up. So, unless we go by the standard model prescription that the neutrino has no mass, we are led to conclude that as noted above the neutrino has a minuscule mass. Now mass is required for neutrino oscillations so herein is the origin of neutrino oscillations referred to above. (Actually, what we know are the squared differences of the neutrino masses.)

3.7   Electrons and Neutrinos

Mathematical Formulations

Several authors have considered negative energy solutions of the Dirac equation and its ramifications. (Cf. Ref. [55] and also [62]) for a review). We now look at this with fresh perspectives and obtain some consequences, which are unanticipated.

At very high energies, negative energies are encountered. The reason is due to the fact that the set of positive-energy solutions of the Dirac or Klein–Gordon equations are not a complete set [59]. At normal energies, the well-known Foldy–Wouthuysen transformation could be applied which yields a description consisting of positive energies alone. That is a description free of operators which mix negative-energy and positive-energy components of the wave function. This picture also leads it to the two-component Pauli equation in the non-relativistic limit [41].

Cini and Toushek have shown independently that when applied to very high energies, the Foldy–Wouthuysen transformation yields a different picture [63]. Let us see this in detail [55].

The Cini–Toushek transformation can be written in the form

$$e^{\pm \iota s} = \frac{E + |p|}{2E} \pm \frac{\vec{\gamma} \cdot \vec{p}}{2E|p|} \cdot m. \tag{3.28}$$

Under (3.28), we know that the Dirac equation goes over to the massless neutrino equation:

$$H\psi = \frac{\vec{\alpha} \cdot \vec{p}}{|p|} E(p). \tag{3.29}$$

Using the following notation:

$$\alpha^k = \begin{pmatrix} 0 & \sigma^k \\ \sigma^k & 0 \end{pmatrix} \qquad \beta = \begin{pmatrix} I & 0 \\ 0 & -I \end{pmatrix} \tag{3.30}$$

$$\gamma^0 = \beta \tag{3.31}$$

$$\gamma^k = \beta \alpha^k \quad (k = 1, 2, 3), \tag{3.32}$$

where σ^k are the Pauli matrices and I is the 2×2 unit matrix.

Also needed here is the γ_5 operator, given by

$$\gamma_5 = \gamma^0 \gamma^1 \gamma^2 \gamma^3 = \iota \begin{pmatrix} I & 0 \\ 0 & -I \end{pmatrix}. \tag{3.33}$$

Using (3.28), the transformed matrix becomes

$$\Gamma_5 = e^{-\iota S} \gamma_5 e^{\iota S} = \left\{ \frac{E + p}{2E} + \frac{(\vec{\gamma} \cdot \vec{n})m}{2E} \right\} \gamma_5 \left\{ \frac{E + p}{2E} - \frac{(\vec{\gamma} \cdot \vec{n})m}{2E} \right\}, \tag{3.34}$$

which finally yields

$$\Gamma_5 = \gamma_5 + \left(\frac{m}{E} \right) (\vec{\gamma} \cdot \vec{n}) \gamma_5, \tag{3.35}$$

where $\vec{n}$ is the unit vector in the direction of the momentum vector. We can see from (3.35) that

$$\Gamma_5 = \gamma_5 \tag{3.36}$$

whenever m is negligible compared to E, $\mathcal{O}\left(\frac{m}{E}\right) << 1$.

The Dirac equation is

$$\left(\gamma^\mu p_\mu - m \right) \psi = 0. \tag{3.37}$$

As commented upon earlier, the two-component spinors belonging to the representation in [55]

$$D^{(\frac{1}{2}0)} \text{ or } D^{(0\frac{1}{2})}$$

of the Lorentz group are solutions of the Dirac equation (3.37). But these can be seen to be no longer invariant under reflection [64]. The 4×4 representation is necessary to maintain invariance under reflection:

$$D^{(\frac{1}{2}0)} \oplus D^{(0\frac{1}{2})}.$$

Under reflection, the two spinors transform into each other. And the overall invariance is left intact [55].

3.8   Spin of the Neutrino

We also noted that, as is known [39], the Maxwell equations can also be written in the form of neutrino equations. Defining a four vector such that

$$\chi_j = E_j + \iota B_j, \quad \chi_0 = 0 \tag{3.38}$$

we can rewrite the Maxwell equations in the form:

$$\beta_\mu \frac{\partial \chi_\nu}{\partial x_\mu} = -\frac{1}{c} j_\nu, \tag{3.39}$$

where, in a particular representation, for example,

$$\beta_0 = I \times I, \quad \beta_1 = -\sigma_3 \otimes \sigma_2,$$

$$\beta_2 = \sigma_2 \otimes I, \quad \beta_3 = \sigma_1 \otimes \sigma_2,$$

"$\times$" refers to the cross product, the σ's being the Pauli matrices and wherein for our source-free vacuum case, the current four vector on the right-hand side of Eq. (3.39) vanishes. It is easy to show that the four-component Eq. (3.39) breaks down into two-component neutrino-like equations, except that both these equations are coupled owing to the additional condition $\chi_0 = 0$ in (3.38). This has been the problem in identifying (3.39) with the Dirac theory.

In the above context, let us now approach the earlier considerations from the opposite point of view that of the Dirac equation. It is well known that the four linearly independent four spinor Dirac wave functions are given by [41], apart from multiplicative factors,

$$\begin{bmatrix} 1 \\ 0 \\ \frac{p_z c}{E+mc^2} \\ \frac{p+c}{E+mc^2} \end{bmatrix} \begin{bmatrix} 0 \\ 1 \\ \frac{p-c}{E+mc^2} \\ \frac{-p_z c}{E+mc^2} \end{bmatrix} \begin{bmatrix} \frac{p_z c}{E+mc^2} \\ \frac{p+c}{E+mc^2} \\ 1 \\ 0 \end{bmatrix} \begin{bmatrix} \frac{p+c}{E+mc^2} \\ \frac{-p_z c}{E+mc^2} \\ 0 \\ 1 \end{bmatrix} \tag{3.40}$$

where p_z is the z-component of the momentum and

$$p_\pm = p_x \pm \iota p_y,$$

in a representation given by

$$\gamma_\iota = \gamma_0 \begin{bmatrix} 0 & \sigma_\iota \\ \sigma_\iota & 0 \end{bmatrix}, \gamma_0 = \begin{bmatrix} 1 & 0 \\ 0 & -1 \end{bmatrix}$$

the σ's being the Pauli matrices. It must be mentioned that any localized solution would contain solutions of both signs of energy.

If we consider the z-axis to be in the direction of motion, for simplicity and take the limit $m \to 0$, the spinors in (3.40) become

$$\psi_1 = \begin{bmatrix} 1 \\ 0 \\ 1 \\ 0 \end{bmatrix} \quad \psi_2 = \begin{bmatrix} 0 \\ 1 \\ 0 \\ -1 \end{bmatrix} \quad \psi_3 = \begin{bmatrix} 1 \\ 0 \\ 1 \\ 0 \end{bmatrix} \quad \psi_4 = \begin{bmatrix} 0 \\ -1 \\ 0 \\ 1 \end{bmatrix}. \tag{3.41}$$

Indeed this is the case for the Cini–Toushek transformation as can be seen from (3.29).

It should be noticed that in (3.41) $\psi_1 = \psi_3$, and $\psi_2 = \psi_4$ so that effectively, two of the spinors vanish exactly and we are left with two solutions as in the case of the solutions χ of (3.39). (The mass zero four-component Dirac spinor does not represent a neutrino unless an auxiliary condition, which effectively destroys the lower two or upper two components, is imposed [55].) It can now be seen from the above considerations that the source-free vacuum electromagnetic field can be in a sense considered to be a composite of a neutrino and an anti-neutrino. For instance, from (3.40), we could see that

$$\begin{bmatrix} 1 \\ 0 \\ 0 \\ 0 \end{bmatrix} + \begin{bmatrix} 0 \\ 0 \\ 0 \\ 1 \end{bmatrix} = \begin{bmatrix} 1 \\ 0 \\ 0 \\ 1 \end{bmatrix}.$$

All this is true in the UHE region where (3.29) would hold.

We must remember that the Eq. (3.39) is actually coupled neutrino equations, coupled by the condition in (3.38). It may be mentioned that the possibility of bosons being bound states of fermions rather than being primary has been discussed by the author and other scholars [30, 37].

3.9 Additional Comments: Early Derivation of the Kerr–Newman Metric

It is interesting that the transformation (3.38) was used by Newman [65, 66] to derive the linearized version of what is today called the Kerr–Newman metric. What was most surprising was that from this purely classical consideration the purely quantum mechanical $g = 2$ emerges. This has been discussed at length by the author [1]: The mystery disappears if we remember that such a complexification of coordinates if generalized to three dimensions leads to a quarternionic description and the four-dimensional Minkowski metric. In effect, we introduce spin half into the formulation.

A slightly different way of looking at this is [67] as follows: Let us start with the charge-free Maxwell equations:

$$div\,E = 0,$$

$$div\,B = 0. \tag{3.42}$$

Scalar multiplication of the curl equations on the right-hand side by the Pauli vector σ

$$\sigma = (\sigma_1, \sigma_2, \sigma_3)$$

$$\sigma_1 = \begin{pmatrix} 0 & 1 \\ 1 & 0 \end{pmatrix}, \sigma_2 = \begin{pmatrix} 0 & -\imath \\ \imath & 0 \end{pmatrix}, \sigma_3 = \begin{pmatrix} 1 & 0 \\ 0 & -1 \end{pmatrix},$$

using the algebraic relation

$$(\sigma \cdot \nabla)(\sigma \cdot A) = div\,A + \imath\sigma \cdot rot\,\mathbf{A} \tag{3.43}$$

together with the two divergence equations (3.42) transforms the system into

$$\left\{ \begin{array}{l} (\sigma \cdot \nabla)(\sigma \cdot .H) - \frac{\varepsilon}{c}\frac{\partial}{\partial t}(\imath\sigma \cdot E) = 0. \\ (\sigma \cdot \nabla)(\sigma \cdot E) + \frac{\mu}{c}\frac{\partial}{\partial t}(\sigma \cdot H) = 0. \end{array} \right\} \tag{3.44}$$

In matrix notation, this reads

$$\left[\begin{pmatrix} 1 & \sigma \\ \sigma & 0 \end{pmatrix} \cdot \nabla - \begin{pmatrix} \varepsilon 1 & 0 \\ 0 & \mu 1 \end{pmatrix} \frac{1}{c}\frac{\partial}{\partial t} \right] \begin{bmatrix} \imath(\sigma \cdot E) \\ (\sigma \cdot H) \end{bmatrix} = 0. \tag{3.45}$$

Denoting the quantity on which the differential operators act by Ψ, that is,

$$\begin{pmatrix} \imath(\sigma \cdot E) \\ (\sigma \cdot H) \end{pmatrix} = \begin{pmatrix} \imath E_3 & \imath(E_1 - E_2) \\ \imath(E_1 + E_2) & -\imath E_3 \\ H_3 & H_1 - \imath H_2 \\ H_1 + \imath H_2 & -H_3 \end{pmatrix} = \begin{pmatrix} \Psi_1^I & \Psi_1^{II} \\ \Psi_2^I & \Psi_2^{II} \\ \Psi_3^I & \Psi_3^{II} \\ \Psi_4^I & \Psi_4^{II} \end{pmatrix} = \Psi \tag{3.46}$$

(the components of the electromagnetic field vectors are denoted by $E = (E_1, E_2, E_3)$ and $H = (H_1, H_2, H_3)$), and considering the well-known connection between the Pauli matrices σ and the Dirac matrices γ:

$$\begin{pmatrix} 0 & \sigma \\ \sigma & 0 \end{pmatrix} = \gamma \tag{3.47}$$

we get for (3.45) the system:

$$\left\{ \left[\gamma \cdot \nabla - \begin{pmatrix} \varepsilon 1 & 0 \\ 0 & \mu 1 \end{pmatrix} \frac{1}{c}\frac{\partial}{\partial t} \right] \Psi = 0 \right\}. \tag{3.48}$$

Here one has to bear in mind that each of the two columns of the matrix (3.46) independently represents a system of functions solving (3.45) or (3.48). From this, a separation of the time dependence according to

$$\Psi = \psi e^{-\iota \omega t} \tag{3.49}$$

finally yields the amplitude equation:

$$\left[\gamma \cdot \nabla + \iota \frac{\omega}{c} \begin{pmatrix} \varepsilon 1 & 0 \\ 0 & \mu 1 \end{pmatrix} \right] \Psi = 0. \tag{3.50}$$

Its agreement with the Dirac equation

$$\left[\gamma \cdot \nabla + \iota \frac{\omega}{c} \begin{pmatrix} \left(1 - \frac{\Phi - m_0 c^2}{\hbar \omega}\right) 1 & 0 \\ 0 & \left(1 - \frac{\Phi + m_0 c^2}{\hbar \omega}\right) 1 \end{pmatrix} \right] \Psi = 0 \tag{3.51}$$

is obvious. That is the multiplication of the charge-free Maxwell equations by the Pauli spin vector leads to the Dirac equation. This is not surprising because spin is a typically quantum mechanical effect.

It may appear strange, how a combination of a neutrino and an anti-neutrino could lead to a spin half electron in the high-energy representation. This can be brought out best by using a formulation due to Feshbach and Villars [59] interpretations. Feshbach and Villars interpreted the KG equation in a single particle rather than field theoretic context. In fact, they showed that this (F-V) formulation also applies to the Dirac equation. To see this, we can rewrite the K-G equation in the Schrodinger form, invoking a two-component wave function:

$$\Psi = \begin{pmatrix} \phi \\ \chi \end{pmatrix}, \tag{3.52}$$

The $K - G$ equation then can be written as (Cf. Ref. [59] for details)

$$\iota \hbar (\partial \phi / \partial t) = (1/2\,m)(\hbar / \iota \nabla - eA/c)^2 (\phi + \chi)$$

$$+ (eA_0 + mc^2)\phi,$$

$$\iota \hbar (\partial \chi / \partial t) = -(1/2\,m)(\hbar / \iota \nabla - eA/c)^2 (\phi + \chi) + (eA_0 - mc^2)\chi \tag{3.53}$$

It will be seen that the components ϕ and χ are coupled in (3.53). In fact, we can analyse this matter further, considering free particle solutions for simplicity. We write

$$\Psi = \begin{pmatrix} \phi_0(p) \\ \chi_0(p) \end{pmatrix} e^{\iota / \hbar (p \cdot x - Et)}$$

$$\Psi = \Psi_0(p) e^{\iota / \hbar (p \cdot x - Et)} \tag{3.54}$$

Introducing (3.54) into (3.53) we obtain, two possible values for the energy E, viz.:

$$E = \pm E_p; \quad E_p = [(cp)^2 + (mc^2)^2]^{\frac{1}{2}}. \tag{3.55}$$

The associated solutions are

$$\begin{aligned} E = E_p \quad &\phi_0^{(+)} = \frac{E_p + mc^2}{2(mc^2 E_p)^{\frac{1}{2}}} \\ \psi_0^{(+)}(p): \quad &\chi_0^{(+)} = \frac{mc^2 - E_p}{2(mc^2 E_p)^{\frac{1}{2}}} \end{aligned} \left. \right\} \ \phi_0^2 - \chi_0^2 = 1,$$

$$\begin{aligned} E = -E_p \quad &\phi_0^{(-)} = \frac{mc^2 - E_p}{2(mc^2 E_p)^{\frac{1}{2}}} \\ \psi_0^{(-)}(p): \quad &\chi_0^{(-)} = \frac{E_p + mc^2}{2(mc^2 E_p)^{\frac{1}{2}}} \end{aligned} \left. \right\} \ \phi_0^2 - \chi_0^2 = -1. \tag{3.56}$$

It can be seen from this that even if we take the positive sign for the energy in (3.55), the ϕ and χ components get interchanged with a sign change for the energy. Furthermore, we can easily show from this that in the non-relativistic limit, the χ component is suppressed by order $(p/mc)^2$ compared to the ϕ component exactly as in the case of the Dirac equation [41].

We now observe that in the above formulation for the wave function

$$\Psi = \begin{pmatrix} \phi \\ \chi \end{pmatrix}, \tag{3.57}$$

where, as noted, ϕ and χ are, for the Dirac equation, each two spinors. ϕ (or more correctly ϕ_0) represents a particle while χ represents an antiparticle. So, for one observer we have

$$\Psi \sim \begin{pmatrix} \phi \\ 0 \end{pmatrix} \tag{3.58}$$

and for another observer we can have

$$\Psi \sim \begin{pmatrix} 0 \\ \chi \end{pmatrix}, \tag{3.59}$$

that is, the two observers would see respectively a particle and an antiparticle. Usually we see localized particles including both (3.58) and (3.59). This would be the same for a single observer, if, for example, the particle's velocity got a boost so that (3.59) rather than (3.58) would dominate after sometime.

It is now easy to see that without any inconsistency or contradiction to the theory, in which case Eq. (3.58) represents a neutrino while Eq. (3.59) represents the antineutrino. However, they are not independent, in the sense that they really describe the Dirac four spinor (3.57), appearing as either (3.58) or (3.59) at different energies, in other words according to the considerations after (3.37) or before (3.38).

The above scenario of antiparticle and particle at different times could be modelled as follows: The very high-energy antiparticle component, $u-$ let us say would decay rapidly, for example, exponentially with time while the normal particle $u+$ would have the usual longer life time. In fact, this is exactly the case with the Kaon decay and the recently observed B-meson decay, in both of which time-reversal symmetry is broken.

3.10 Transmutations of Particles

We would like to point out two things with regard to Eqs. (3.57), (3.58), (3.59). The first is that the transition from (3.58) to (3.59), that is, particle to antiparticle or vice versa with a steep increase in energy can be looked upon as a particle–antiparticle "transmutation". Indeed such "transmutations" have been observed recently in the so-called B-factories involving the decays of B-mesons. Secondly (3.58) and (3.59) in a previous communication [62] have been described as the up and down states of a super spin—that is, particles and antiparticles rather like two different states of the same entity given by (3.57).

3.11 Neutrino *Waves*

Recently the 60-year-old Glashow resonance was detected in Antarctica in the ice cube experiment. Sniffing out such a minute neutrino effect is indeed an awesome achievement. This was verified by the Russians in Lake Baikal in Siberia. In this spirit, we now propose what may be called neutrino "waves".

Our starting point is from an infinitesimal parallel displacement of a vector, namely:

$$\delta a^\sigma = \Gamma^\sigma_{\mu\nu} a^\mu dx^\nu. \tag{3.60}$$

This represents the displacement effects due to the curvature of space [30]. If space were flat, the right side of (3.60) would be zero. Considering the μ coordinate, we get

$$\frac{\partial a^\sigma}{\partial x^\mu} \rightarrow \frac{\partial a^\sigma}{\partial x^\mu} - \Gamma^\sigma_{\mu\nu} a^\nu \tag{3.61}$$

which would also be written as

$$- \Gamma^\lambda_{\mu\nu} g^\nu_\lambda a^\sigma = -\Gamma^\nu_{\mu\nu} a^\sigma.$$

Now let us write the metric as

$$g_{\mu\nu} = \eta_{\mu\nu} + h_{\mu\nu}. \tag{3.62}$$

In the above (3.62) $\eta_{\mu\nu}$ is the Minkowski metric and $h_{\mu\nu}$ would be a very small perturbative effect. So we get

$$\frac{\partial}{\partial x^\mu} \rightarrow \frac{\partial}{\partial x^\mu} - \Gamma^\nu_{\mu\nu}. \tag{3.63}$$

This is very similar to the minimum electromagnetic coupling and the last term in (3.63) would be the interaction term. However it must be remembered that even though the neutrino wave function can be represented in the usual manner [41], in terms of column vectors, in the above context let us now approach the above considerations of that of the Dirac equation. It is well known that the four linearly independent four spinor Dirac wave functions are given by [41], apart from multiplicative factors,

$$\begin{bmatrix} 1 \\ 0 \\ \frac{p_z c}{E+mc^2} \\ \frac{p+c}{E+mc^2} \end{bmatrix} \begin{bmatrix} 0 \\ 1 \\ \frac{p-c}{E+mc^2} \\ \frac{-p_z c}{E+mc^2} \end{bmatrix} \begin{bmatrix} \frac{p_z c}{E+mc^2} \\ \frac{p+c}{E+mc^2} \\ 1 \\ 0 \end{bmatrix} \begin{bmatrix} \frac{p+c}{E+mc^2} \\ \frac{-p_z c}{E+mc^2} \\ 0 \\ 1 \end{bmatrix}, \tag{3.64}$$

where p_z is the z-component of the momentum and

$$p_\pm = p_x \pm \iota p_y,$$

in a representation given by

$$\gamma_\iota = \gamma_0 \begin{bmatrix} 0 & \sigma_\iota \\ \sigma_\iota & 0 \end{bmatrix}, \gamma_0 = \begin{bmatrix} 1 & 0 \\ 0 & -1 \end{bmatrix}$$

the σ's being the Pauli matrices.

If we consider the z-axis to be in the direction of motion, for simplicity and taking the limit $m \rightarrow 0$, the spinors become

$$\psi_1 = \begin{bmatrix} 1 \\ 0 \\ 1 \\ 0 \end{bmatrix} \psi_2 = \begin{bmatrix} 0 \\ 1 \\ 0 \\ -1 \end{bmatrix} \psi_3 = \begin{bmatrix} 1 \\ 0 \\ 1 \\ 0 \end{bmatrix} \psi_4 = \begin{bmatrix} 0 \\ -1 \\ 0 \\ 1 \end{bmatrix}. \tag{3.65}$$

It should be noticed that here $\psi_1 = \psi_3$, and $\psi_2 = \psi_4$ so that effectively, two of the spinors vanish exactly and we are left with two solutions. This was elaborated upon by several authors starting from Dirac [42], Barut [39], and Sidharth [14].

We now observe that in the linearized theory of general relativity [68–70], we have

$$\Box h_{\mu\nu} = -16\pi T_{\mu\nu}. \tag{3.66}$$

This is a wave equation á la linearized gravity waves which have been elaborated upon by the author [14]. Thus these are extremely faint waves or perturbations. Nevertheless it may now be possible to find them.

As noted, the neutrino was introduced by Wolfgang Pauli. To balance the neutron decay equation

$$n \rightarrow p^+ e^- \bar{\nu}_e.$$

The neutron decays into a proton, an electron, and an anti-neutrino of the electron type. For the balance to happen, the neutrino had to be of spin half and mass zero. That is how it has been in the standard model of particle physics. But over the past 50 years the neutrino, if anything has become even more enigmatic and studies are continuing. Let us consider the neutrino by the usual rules of scattering. We get in the special case, where the mass is considered to be nearly zero, the formula for scattering becoming that of the Fermi point interaction. This is a clear verdict on the weakness of the neutrino interactions.

We next consider the neutrino beam which is obviously mono-energetic. Such beams have been investigated by the author before 2008. The conclusion was that these beams are two dimensional in momentum space and therefore also in ordinary space time as seen earlier [14].

Let us now specialize to two-dimensional fermions. It can be argued that they behave like mono-atomic particles. A quick check of this is the following argument: In the case of mono-atomic particles, as also neutrinos in 2D we have for the pressure, volume, and energy, the relation

$$PV = \frac{2}{3}E. \tag{3.67}$$

Let us now return to such a collection, in this case, by using ordinary statistics. We have, as is well known, the number of particles in the interval $(c, c + dc)$ is given by $dN = A \cdot N$, A being a constant. The point is that there is a non-zero probability for the particles having a velocity $> c$.

It may be mentioned that superluminal neutrinos were observed in the supernova 1987A. On the other hand, the Gran Sasso underground neutrino experiment which supposedly threw up superluminal neutrinos turned out to be due to a faulty electrical connection. In any case, we would like to say that a few superluminal neutrinos may be observed with a very small probability.

We discuss further ways in which we can observe superluminosity in neutrinos, albeit in the probabilistic fashion. The first way is by considering a stream of mono-energetic neutrinos. This leads to two dimensionality as we saw earlier.

The other is that if we leave the neutrino to explore the Weinberg spacetime interval, [71] there is a small gap which can be used for particles to go through. However, the effect would be minuscule except in the case of the neutrino which has a large Compton wave length. Finally, we showed that neutrinos are like waves, which is another way of saying that they are spread out. The question is: exactly at which point is the neutrino located. When quantum mechanical effects are factored in, we have a modification of the usual formula. Specifically, the uncertainty principle tells us that when we specify that a particle is at position x_1 at time t_1, we cannot also define its velocity precisely. In consequence there is a certain chance of a particle getting from x_1 to x_2 even if $x_1 - x_2$ is spacelike. That is,

$$|x_1 - x_2| > |x_1^0 - x_2^0|. \tag{3.68}$$

To be more precise, the probability of a particle reaching x_2 if it starts at x_1 is nonnegligible as long as

$$(x_1 - x_2)^2 - (x_1^0 - x_2^0)^2 \leq \frac{\hbar^2}{m^2}, \tag{3.69}$$

where $\hbar$ is Planck's constant (divided by 2π) and m is the particle mass. (Such spacetime intervals are very small even for elementary particle masses; for instance, if m is the mass of a proton then $\hbar/m = 2 \times 10^{-14} cm$ or in time units $6 \times 10^{-25} sec$. Recall that in our units $1sec = 3 \times 10^{10} cm$.) There is a small difference, which from Eq. (3.69) is negligible, except in the case of neutrinos which have zero mass or the slightest of masses. Within this extended interval, the neutrino could have a superluminal speed.

Remark on Superluminosity

If we consider an electron collision with a neutrino as noted, we get the Fermi point interaction represented by an X in propagator language [41].

We finally saw that there is one small chink through which Einstein's relativity can be hoodwinked.

Coming to the faster than light question, it turns out that due to quantum mechanics, what is called the spacetime interval gets slightly extended, just by a wee bit, as we saw, which gives the opportunity to move faster than light. This wee bit is the maximum for particles with the tiniest of masses, namely, the neutrinos. We have written quite a bit on what may be called superluminal or faster than light neutrinos. As noted, the 1987A supernova detection is a little more serious and awaits a jury. In any case, it is conceivable that we can use superluminal neutrinos for distant communications.

3.12   Is the Neutrino a Hybrid Particle?

In this case, the earlier comments lead us to expect bosonic behaviour because there is hardly any energy spread. Our starting point is the well-known formula for the occupation number of a fermion gas [20]:

$$\bar{n}_p = \frac{1}{z^{-1}e^{bE_p} + 1},\tag{3.70}$$

where $z' \equiv \frac{\lambda^3}{v} \equiv \mu z \approx z$ because, here, as can be easily shown $\mu \approx 1$,

$$v = \frac{V}{N}, \lambda = \sqrt{\frac{2\pi\hbar^2}{m/b}}$$

$$b \equiv \left(\frac{1}{KT}\right), \quad \text{and} \quad \sum \bar{n}_p = N.\tag{3.71}$$

Let us consider, in particular, a collection of fermions which is somehow made nearly mono-energetic, that is, given by the distribution:

$$n'_p = \delta(\vec{p} - p_0)\bar{n}_p\tag{3.72}$$

p_0 being the magnitude of the 0th component of the momentum vector $\vec{p}$ where $\bar{n}_p$ is given by (3.70).

This is not possible in general—here we consider a special situation of a collection of mono-energetic particles in equilibrium which is the idealization of a contrived experimental setup.

By the usual formulation, we have

$$N = \frac{V}{\hbar^3}\int d\vec{p}n'_p = \frac{V}{\hbar^3}\int \delta(\vec{p} - p_0)4\pi\,\vec{p}^2\bar{n}_p d\vec{p} = \frac{4\pi V}{\hbar^3}p_0^2\frac{1}{z^{-1}e^\theta + 1},\tag{3.73}$$

where $\theta \equiv bE_{p_0}$.

It must be noted that in (3.73) there is a loss of dimension in momentum space, due to the δ function in (3.72)—in fact, such a fractal two-dimensional situation would be in the relativistic case lead us back to the anomalous behaviour already alluded to [72]. This again is symptomatic of distances in space (and momentum space) being more a measure of dispersion, rather than rigid distances. In the non-relativistic case, two dimensions would imply that the coordinate ψ of the spherical polar coordinates (r, ψ, ϕ) would become constant, $\pi/2$ in fact. In this case, the usual quantum numbers l and m of the spherical harmonics [73] no longer play a role in the usual radial wave equation

$$\frac{d^2u}{dr^2} + \left\{\frac{2m}{\hbar^2}[E - V(r)] - \frac{l(l+1)}{r^2}\right\}u - 0.\tag{3.74}$$

The coefficient of the centrifugal term $l(l + 1)$ in (3.74) is replaced by m^2 as in classical theory [74].

To proceed, in this case, $KT - \!\!< E_p >\approx E_p$ so that $\theta \approx 1$. But we can continue without giving θ any specific value.

Using the expressions for v and z given in (3.71) in (3.72), we get

$$(z^{-1}e^{\theta} + 1) = (4\pi)^{5/2}\frac{z'^{-1}}{p_0}, \text{ whence}$$

$$z'^{-1}A \equiv z'^{-1}\left(\frac{(4\pi)^{5/2}}{p_0} - e^{\theta}\right) = 1, \tag{3.75}$$

where we use the fact that in (3.71), $\mu \approx 1$ as can be easily deduced.

A number of conclusions can be drawn from (3.75). For example, if

$$A \approx 1, i.e.$$

$$p_0 \approx \frac{(4\pi)^{5/2}}{1+e}, \tag{3.76}$$

where A is given in (3.75), then $z' \approx 1$. Remembering that in (3.71), λ is of the order of the de Broglie wavelength and v is the average volume occupied per particle, this means that the gas gets very densely packed for momenta given by (3.76). In fact for a Bose gas, as is well known, this is the condition for Bose–Einstein condensation at the level $p = 0$ (cf. Ref. [20]).

On the other hand, if

$$A \approx 0(\text{that is } \frac{(4\pi)^{5/2}}{e} \approx p_0)$$

then $z' \approx 0$. That is, the gas becomes dilute or V increases.

More generally, Eq. (3.75) also puts a restriction on the energy (or momentum), because $z' > 0$, viz.:

$$A > 0(i.e.p_0 < \frac{(4\pi)^{5/2}}{e})$$

$$\text{But if } A < 0, (i.e.p_0 > \frac{(4\pi)^{5/2}}{e}),$$

then there is an apparent contradiction.

The contradiction disappears if we realize that $A \approx 0$ or

$$p_0 = \frac{(4\pi)^{5/2}}{e} \tag{3.77}$$

(corresponding to a temperature given by $KT = \frac{p_0^2}{2m}$) is a threshold momentum (phase transition). For momenta greater than the threshold given by (3.77), the collection of fermions behaves like bosons. In this case, the occupation number is given by

$$\bar{n}_p = \frac{1}{z^{-1}e^{bE_p} - 1}$$

instead of (3.70), and the right-side equation of (3.75) would be given by '-1' instead of $+1$, so that there would be no contradiction.

References

1. B.G. Sidharth. *The Universe of Fluctuations* (Springer, 2005)
2. S. Perlmutter, Discovery of a supernova explosion at half the age of the Universe. Nature **391**(6662), 51–54 (1998)
3. T. Padmanabhan, Accelerated expansion of the universe driven by tachyonic matter. Phys. Rev. D **66**, 021301 (2002). (Jun)
4. M. Sami, P. Chingangbam, T. Qureshi, Aspects of tachyonic inflation with an exponential potential. Phys. Rev. D **66**(4), 043530 (2002)
5. R.L. Abramo, F. Fabio, Cosmological dynamics of the tachyon with an inverse power-law potential. Phys. Lett. B **575**(3), 165–171 (2003)
6. A. de la Macorra, Interacting dark energy: decay into fermions. Astropart. Phys. **28**(2), 196–204 (2007)
7. P.S. Corasaniti, M. Kunz, D. Parkinson, E.J. Copeland, B.A. Bassett, Foundations of observing Dark Energy dynamics with the Wilkinson microwave anisotropy probe. Phys. Rev. D **70**, 083006 (2004). (Oct)
8. S. Nojiri, S.D. Odintsov, The new form of the equation of state for Dark Energy fluid and accelerating universe. Phys. Lett. B **639**(3–4), 144–150 (2006)
9. M. Szydłowski, O. Hrycyna, Dissipative or conservative cosmology with Dark Energy? Ann. Phys. **322**(12), 2745–2775 (2007)
10. G. Izquierdo, D. Pavón, Dark Energy and the generalized second law. Phys. Lett. B **633**(4–5), 420–426 (2006)
11. N. Straumann, Dark energy: recent developments. Mod. Phys. Lett. A **21**(14), 1083–1097 (2006)
12. U. França, Dark energy, curvature, and cosmic coincidence. Phys. Lett. B **641**(5), 351–356 (2006)
13. W. Zhao, Y. Zhang, Coincidence problem in YM field Dark Energy model. Phys. Lett. B **640**(3), 69 73 (2006)
14. B.G. Sidharth, *The Thermodynamic Universe* (World Scientific, Singapore, 2008)
15. B.G. Sidharth, A model for neutrinos. Int. J. Theor. Phys. **52**, 4412–4415 (2013)
16. B.G. Sidharth, Neutrinos and the weak interactions. Found. Phys. Lett. **18**(4), 393–399 (2005)
17. S. Hayakawa, Atomism and cosmology. Prog. Theor. Phys. Suppl. **65**, 532–541 (1965)
18. B.G. Sidharth, A note on the cosmic neutrino background and the cosmological constant. Found. Phys. Lett. **19**(7), 757–759 (2006)
19. B.G. Sidharth. A Note on the Mass of the Neutrino. arXiv preprint hep-th/9807075 (1998)
20. K. Huang, *Statistical Mechanics* (Wiley Eastern, New Delhi, 1975)
21. L. Fang, R. Ruffini (eds.), *Cosmology of the Early Universe.* Advanced Series in Astrophysics and Cosmology. (World Scientific, Singapore, 1984)
22. M. Tegmark et al., Cosmological parameters from SDSS and WMAP. Phys. Rev. D **69**, 103501 (2004). (May)
23. R. Trotta, A. Melchiorri, Indication for primordial anisotropies in the neutrino background from the Wilkinson microwave anisotropy probe and the Sloan digital sky survey. Phys. Rev. Lett. **95**, 011305 (2005). (Jun)
24. T.J. Weiler, Cosmic neutrino physics and astrophysics. Int. J. Mod. Phys. A **20**(06), 1168–1179 (2005)

25. S. Hayakawa, Cosmological interpretation of the weak forces. Prog. Theoret. Phys. **33**(3), 538–539 (1965)
26. J.D. Barrow, D.J. Shaw, The value of the cosmological constant. Int. J. Mod. Phys. D **20**(14), 2875–2880 (2011)
27. B.G. Sidharth, A note on the cosmic neutrino background and the cosmological constant. Found. Phys. Lett. **19**(7), 757–759 (2006)
28. S. Weinberg, The cosmological constant problem. Rev. Mod. Phys. **61**, 1–23 (1989). (Jan)
29. B.G. Sidharth, Issues in quantized fractal spacetime. Chaos, Solitons & Fractals **12**(8), 1449–1457 (2001)
30. B.G. Sidharth, *Chaotic Universe* (Nova Science, New York, 2001)
31. B.G. Sidharth, Different routes to Lorentz symmetry violations. Found. Phys. **38**(1), 89–95 (2008)
32. Y. Adiv, H. Hu, S. Tsesses, R. Dahan, K. Wang et al., Observation of 2d Cherenkov radiation. Phys. Rev. X **13**(1), 011002 (2023)
33. B.G. Sidharth, Noncommutative spacetime, mass generation, and other effects. Int. J. Mod. Phys. E **19**(01), 79–90 (2010)
34. B.G. Sidharth, The mass of the neutrinos. Int. J. Mod. Phys. E **19**(11), 2285–2292 (2010)
35. J.A. Formaggio, C.A. Luiz, R.G.H. Robertson, Direct measurements of neutrino mass. Phys. Rep. **914**, 1–54 (2021). Direct measurements of neutrino mass
36. R.N. Mohapatra, A.Y. Smirnov, Neutrino Mass and New Physics. Annu. Rev. Nucl. Part. Sci. **56**(1), 569–628 (2006)
37. J.R. Klauder, Bosons Without Bosons, in *Conference on Quantum Theory and the Structures of Time and Space 3*, number PRINT-79-0016 (BTL) (1978)
38. D. Bazeia, A. Mohammadi, D.C. Moreira, Fermion bound states in geometrically deformed backgrounds. Chin. Phys. C **43**(1), 013101 (2019). (Jan)
39. A.O. Barut, *Electrodynamics and Classical Theory of Fields and Particles* (1980)
40. B.G. Sidharth, Low Dimensional Electrons. *Solid State Physics, Eds. R. Mukhopadhyay et. al*, 41:331 (1999)
41. J.D. Bjorken, S.D. Drell, Relativistic Quantum Mechanics (Mcgraw-Hill College, 1964)
42. P.A.M. Dirac, *The Principles of Quantum Mechanics*, vol. 27 (Oxford university press, 1981)
43. B.G. Sidharth, A report on gravitation as a Planck scale effect. New Adv. Phys. **5**(2), 55–56 (2011)
44. B.G. Sidharth, The modified Dirac equation. Electron. J. Theor. Phys. **7**(24), 211–218 (2010)
45. N. Jachowicz, A. Nikolakopoulos, Nuclear medium effects in neutrino-and antineutrino-nucleus scattering. Eur. Phys. J. Special Topics **230**(24), 4339–4356 (2021)
46. J.A. Formaggio, Neutrino Experiments come closer to seeing CP Violation. Physics **7**, 15 (2014)
47. K. Abe, J. Adam et al., Observation of electron neutrino appearance in a muon neutrino beam. Phys. Rev. Lett. **112**, 061802 (2014). (Feb)
48. S. Rainville et al., A direct test of $e = mc^2$. Nature **438**(7071), 1096–1097 (2005)
49. C.W. Misner, K.S. Thorne, *and J* (Freeman, A. Wheeler. Gravitation, 2000)
50. B.G. Sidharth, Anomalous Fermions. J. Stat. Phys. **95**(3), 775–784 (1999)
51. B.G. Sidharth, A note on Two Dimensional Fermions. *BSC-CAMCS-TR-95-04-01* (1995)
52. A.K. Geim, Graphene: status and prospects. Science **324**(5934), 1530–1534 (2009)
53. A.H. Castro Neto, F. Guinea, N.M.R. Peres, K.S. Novoselov, A.K. Geim, The electronic properties of Graphene. Rev. Mod. Phys. **81**, 109–162 (2009)
54. K.S. Novoselov, A.K. Geim et al., Two-dimensional gas of massless Dirac fermions in Graphene. Nature **438**(7065), 197–200 (2005)
55. S.S. Schweber, *An introduction to relativistic Quantum Field Theory* (Courier Corporation, 2011)
56. B.G. Sidharth, Issues in quantized fractal spacetime. Chaos, Solitons & Fractals **12**(8), 1449–1457 (2001)
57. E. Witten, Anti-de Sitter space, thermal phase transition, and confinement in gauge theories. Adv. Theor. Math. Phys. **2**(3), 505–532 (1998)

58. A.D. Popova, B.G. Sidharth, Is the Universe asymptotically spatially two dimensional. Nonlinear World **4**, 493–504 (1997)
59. H. Feshbach, F. Villars, Elementary relativistic wave mechanics of spin 0 and spin 1/2 particles. Rev. Mod. Phys. **30**(1), 24–25 (1958)
60. W. Greiner, B. Muller, J. Rafelski, *Quantum Electrodynamics of Strong Fields* (Springer, 1985)
61. B.G. Sidharth, Modeling Dark energy and the Higgs Boson. New Adv. Phys. **6**(2), 71–75 (2012)
62. B.G. Sidharth, Negative energy solutions and Symmetries. Int. J. Mod. Phys. E **20**(10), 2177–2188 (2011)
63. M. Cini, B. Touschek, The relativistic limit of the theory of spin 1/2 particles. Il Nuovo Cimento **7**(3), 422–423 (1958)
64. V. Heine, *Group TheoryiIn Quantum Mechanics: An Introduction to Its Present Usage* (Courier Corporation, 2007)
65. E.T. Newman, Maxwell's equations and complex Minkowski space. J. Math. Phys. **14**(1), 102–103 (1973)
66. E.T. Newman, Complex coordinate transformations and the Schwarzschild-Kerr metrics. J. Math. Phys. **14**(6), 774–776 (1973)
67. H.H. Salhofer, *Essays on the Formal Aspects of Electromagnetic Theory*, ed. by A. Lakhtakia (World Scientific, 1993)
68. S. Aksteiner, L. Andersson, Linearized gravity and gauge conditions. Class. Quantum Gravity **28**(6), 065001 (2011)
69. C.J. Fewster, D.S. Hunt, Quantization of linearized gravity in cosmological vacuum spacetimes. Rev. Math. Phys. **25**(02), 1330003 (2013)
70. S. Aksteiner, L. Andersson, T. Bäckdahl, New identities for linearized gravity on the Kerr spacetime. Phys. Rev. D **99**(4), 044043 (2019)
71. S. Weinberg, *Gravitation and Cosmology: Principles and Applications of the General Theory of Relativity* (Wiley, 1972)
72. B.G. Sidharth, Interstellar Hydrogen and cosmic background radiation. Chaos, Solitons & Fractals **12**(8), 1563–1564 (2001)
73. J.L. Powell, B. Crasemann, *Quantum Mechanics* (Courier Dover Publications, 2015)
74. H. Goldstein, *Classical Mechanics*. American Association of Physics Teachers (2002)

Chapter 4
The Bizarre Spacetime

4.1 Why Noncommutative Geometry Puts a Bound on Velocities?

The noncommutative feature of spacetime geometry is a topic of great interest. There is a vast amount of literature existing on this subject [1–4]. Particularly, the author has used noncommutativity to provide a feasible interpretation for several phenomena [5, 6]. The objective of the current chapter is to further explore this unique and intrinsic nature of spacetime on the premise of the noncommutativity introduced by Snyder [7] rather than to the Moyal–Weyl formalism [3, 8].

Particularly, we apply this noncommutativity to the Klein-Gordon equation and modify it considering the Compton length to be the fundamental length. On the other hand, for Snyder, it was a general small length. But as pointed out by Snyder himself the idea that a modification of the ordinary concept of spacetime was necessary because the "elementary" particles had fixed masses and associated Compton wavelengths. In other words, it is assumed that the spectra of the spacetime coordinate operators were invariant under Lorentz transformations. The principal result in Snyder's work was that there exists a Lorentz invariant spacetime in which there is a natural unit of length. The introduction of such a unit of length would remove many of the divergence troubles of field theory. The concept of noncommutativity of x, y, z, and t, would follow in a natural manner by the introduction of a smallest unit of length in spacetime according to Snyder's analysis, otherwise the assumption of Lorentz invariance of the spectra of the operators x, y, z, and t, if they commute, would imply continuous spectra.

The second section deals with the modification of the Klein–Gordon equation by involving a parameter representing the noncommutative feature. In the third section, we modify the Foldy–Wouthuysen and the Cini–Toushek transformations that represent the low-energy and high-energy scenarios, respectively. In the fourth section, we investigate further regarding the parameter referred to as noncommutativity and in that course we find some novel results concerning the Lorentz factor.

B. G. Sidharth, *The Dark Energy Paradigm*,
https://doi.org/10.1007/978-981-96-3745-4_4

4.2 Modified Klein–Gordon Equation

As the author has mentioned in several papers [5] and several references therein, the consideration of complex time $i \times t$ leads to the Minkowski spacetime formalism while the ordinary time coordinate (t) leads to the compact four space representing the *zitterbewegung* region. This effect was noticed by Dirac himself when he came up with his equation: the rapidly oscillating solutions were apparently unphysical. Dirac's explanation was that our physical measurements are never instantaneous but rather spread over a small interval—it turns out to be the Compton time [9]. *Zitterbewegung* has been studied a lot over the years, notably by Huang, Hestenes, Kaiser, and other scholars including the author himself [10–12]. More recently the author has re-examined it in the light of the Feshbach–Villars formulation [13]. To put it simply the four-component Dirac wave function can be written as

$$\psi = \begin{pmatrix} \chi \\ \phi \end{pmatrix},$$

where χ and ϕ are each two-component spinors. χ are the so-called "high-energy" spinors and ϕ the low-energy ones. The former become pronounced at high energies and the latter at lower energies. So the Dirac four spinor divides spacetime into two broad regions—the high-energy region where χ dominates and the usual low-energy region where ϕ dominates. We begin with the following complexified identification of time:

$$t \longmapsto \alpha t + \beta i t', \tag{4.1}$$

where α and β are parameters that represent the scale below and above the Compton length, respectively. When one considers phenomena below the Compton scale, $\beta = 0$, and have the ordinary time coordinate, t. Again, when one considers phenomena above the Compton scale, $\alpha = 0$, we have the complex time coordinate, it'. Thus, relation (4.1) represents a region which is the juncture between the compact four space and the non-compact Minkwoski space. The metrics for the two different regions are, respectively:

$$x^2 + y^2 + z^2 + c^2 t^2$$

and

$$x^2 + y^2 + z^2 - c^2 t'^2.$$

Now, for the juncture region, we write

$$\mathrm{d}s^2 = x^2 + y^2 + z^2 + tt'. \tag{4.2}$$

This yields

$$\mathrm{d}s^2 = x^2 + y^2 + z^2 + \alpha c^2 t^2 - \beta c^2 t'^2. \tag{4.3}$$

This represents the critical region that is the boundary of the two regions. Ostensibly, this is the region of the Compton length where the noncommutative nature of spacetime comes into play. Now, the d'Alembertian operator is

$$\frac{\partial^2}{\partial x_\mu \partial x^\nu} = \nabla^2 - \frac{1}{c^2}\frac{\partial^2}{\partial t^2},$$

where, using the noncommutative feature given in (4.4) and (4.5), the different indices μ and ν lead to the d'Alembertian using Snyder's noncommutativity relation [7]. From the following fundamental relation of noncommutativity due to Snyder [7], we have

$$[x_\mu, x_\nu] = \eta\beta(l^2), \tag{4.4}$$

where $\beta(l^2) = \beta_{\mu\nu}(l^2)$ is a real, nonsingular, and antisymmetric square matrix and l is the Compton length. Since this relation is valid for any length segment it is valid for differentials also such as

$$[dx_\mu, dx_\nu] = \xi\beta(l^2) \tag{4.5}$$

because the differentials measure infinitesimal lengths and should concur with relation (4.4). It must be re-emphasized that Snyder's original work leading to (4.4) was based on the existence of a minimum measurable length in a classical context. Throughout we consider the special case of the minimum length to be the Compton length. So (4.5) would also be valid for differentials which are small lengths. This was emphasized in an earlier communication [14]. Therefore, we have

$$dx_\mu dx_\nu = dx_\nu dx_\mu + \xi\beta(l^2).$$

From this relation, we may write

$$\frac{d^2}{dx_\mu dx^\nu} = \frac{d^2}{dx^\nu dx_\mu}\left[\frac{1}{1 + \frac{\xi\beta(l^2)}{dx^\nu dx_\mu}}\right].$$

Now, since $\beta(l^2)$ is very small due to the Compton length "l" being very small, the above relation can be approximated as

$$\frac{d^2}{dx_\mu dx^\nu} = \left[1 - \frac{\xi\beta(l^2)}{dx^\nu dx_\mu}\right]\frac{d^2}{dx^\nu dx_\mu}. \tag{4.6}$$

In terms of partial derivatives, we have

$$\frac{\partial^2}{\partial x_\mu \partial x^\nu} = \left[1 - \frac{\xi\beta(l^2)}{\partial x^\nu \partial x_\mu}\right]\frac{\partial^2}{\partial x^\nu \partial x_\mu}, \tag{4.7}$$

where $\beta(l^2)$ bears the signature of a noncommutative spacetime. This has also been shown elsewhere [15]. Anyway, we shall use relation (4.7) which is essentially a transformation on account of noncommutativity, in the case of the Klein–Gordon equation which can be written as

$$\frac{\partial^2 \psi}{\partial x_\mu \partial x^\nu} + \mu^2 \psi = 0, \tag{4.8}$$

where $\psi = \psi(x, t)$ is the wave function and $\mu\ (= \frac{mc}{\hbar})$ is the mass-related term. Again, considering the transformation relation (4.4) for the d'Alembertian, using the noncommutative relations, and interchanging the indices ($\mu \leftrightarrow \nu$) we can also write for (4.8)

$$\left[1 - \frac{\eta\beta(l^2)}{\delta x_\nu \delta x_\mu}\right] \frac{\partial^2 \psi}{\partial x_\mu \partial x^\nu} = \mu^2 \psi. \tag{4.9}$$

Now, writing

$$\left[1 - \frac{\eta\beta(l^2)}{\delta x_\nu \delta x_\mu}\right] = \zeta$$

and

$$\mu' = \frac{\mu}{\zeta}$$

we have finally the modified Klein–Gordon equation as

$$\frac{\partial^2 \psi}{\partial x_\mu \partial x^\nu} + \mu'^2 \psi = 0. \tag{4.10}$$

This is like Eq. (4.8). Here, ζ is nearly equal to 1, since the $\beta(l^2)$ is infinitesimal. It is almost as if the massless Goldstone Bosons which satisfy Eq. (4.8) acquire a miniscule mass because of the noncommutativity. Equation (4.10) relates the noncommutative feature of spacetime through the matrix $\beta(l^2)$ with the mass parameter (μ). One can infer that the generation of mass is due to the noncommutative spacetime. This is because we see from Eq. (4.9), $[1 - \frac{\varepsilon\beta(l^2)}{\delta x^\nu \delta x_\mu}]\frac{\partial^2}{\partial x_\mu \partial x^\nu}$ is an operator that operates on the wave function and produces the mass-related term μ. This shows that the noncommutative feature of spacetime can be indeed interesting when taken into consideration. We shall see this more in the subsequent sections where we obtain a slight modification of the energy levels owing to the parameter ζ. This parameter, as we shall see, leads to results that are in good accord with some observational phenomena. It appears that, if the noncommutative nature of spacetime is neglected, then the parameter $\beta(l^2)$ is 0, which leads to $\zeta = 1$ and we have the usual Klein–Gordon formulation.

4.3   The Modified Transformations for High-Energy and Low-Energy Scenarios

In the previous section, we have derived a modified form of the Klein–Gordon equation, the modification itself arising from noncommutativity. Here, we shall derive modified forms of the Foldy–Wouthuysen [16–19] and the Cini–Toushek transformations [18, 20]. Now, the modified Klein–Gordon equation (4.10) can also be written as

$$(\partial_\mu \partial_\nu + \mu'^2)\psi = 0. \tag{4.11}$$

This can also be written as

$$(i\gamma_\mu \partial_\mu + \mu')(-i\gamma_\nu \partial_\nu + \mu')\psi = 0. \tag{4.12}$$

From (4.12), as we know, one can infer

$$(\gamma_\mu p_\mu + \mu')\psi = 0$$

or

$$(-\gamma_\mu p_\mu + \mu')\psi = 0,$$

where we have chosen $\hbar = c = 1$ and $p_\mu = i\partial_\mu$ and the $\gamma's$ are Dirac matrices. Again, in the presence of an electromagnetic interaction, these two equations can be rewritten as

$$[\gamma_\mu(p_\mu - eA_\mu) + \mu']\psi = 0 \tag{4.13}$$

and

$$[\gamma_\mu(-p_\mu + eA_\mu) + \mu']\psi = 0. \tag{4.14}$$

It is obvious that if Eq. (4.13) represents a particle of mass "m" and charge "e" then Eq. (4.14) represents the antiparticle with mass "m" and charge "-e". Therefore, as we see the two equations that derive from the Klein–Gordon Equation (4.11) correspond to a matter–antimatter asymmetry. Now, without any interaction the equations can also be written, respectively, as

$$(\alpha \cdot \vec{p} + \beta \frac{m}{\zeta_1})\psi = \iota \frac{\partial \psi}{\partial t} \tag{4.15}$$

and

$$(\alpha \cdot \vec{p} - \beta \frac{m}{\zeta_2})\psi = i \frac{\partial \psi}{\partial t}, \tag{4.16}$$

where α and β are the usual matrices. Here, apparently we have distinguished between the ζ's for the two equations (4.15) and (4.16). The rationale for this is the fact that these two distinct equations represent a particle and an antiparticle. Interestingly, we shall see later that the former corresponds to the low-energy case and the latter to the high-energy case. Now, let us consider a unitary transformation for Eq. (4.15)

$$U = e^{is},$$

$$\psi' = e^{is}\psi$$

such that we have

$$i\frac{\partial\psi}{\partial t} = e^{is}H\psi$$
$$= e^{is}He^{-is}\psi'$$
$$= H'\psi',$$

where H is the usual Dirac Hamiltonian. As we know [16], such a choice of transformation is given by

$$e^{is} = e^{(\beta\alpha\cdot\vec{p}\theta(\vec{p}))} = \cos p\theta + \beta\frac{\alpha\cdot\vec{p}}{p}\sin p\theta. \tag{4.17}$$

Thus, the transformed Hamiltonian is given as

$$
\begin{aligned}
H' &= \left(\cos p\theta + \beta\frac{\alpha\cdot\vec{p}}{p}\sin p\theta\right)\left((\alpha\cdot\vec{p}) + \beta\frac{m}{\zeta_1}\right)\left(\cos p\theta - \beta\frac{\alpha\cdot\vec{p}}{p}\sin p\theta\right) \\
&= \alpha\cdot\vec{p}\left(\cos 2p\theta - \beta\frac{m}{\zeta_1 p}\sin 2p\theta\right) + \beta\left(\frac{m}{\zeta_1}\cos 2p\theta + p\sin 2p\theta\right).
\end{aligned}
$$

Putting

$$\tan 2p\theta = \frac{\zeta_1 p}{m} \tag{4.18}$$

in the first term of the last line, we obtain

$$H' = \beta\left(\frac{m}{\zeta_1}\cos 2p\theta + p\sin 2p\theta\right).$$

Now, squaring both sides, using (4.18) and after some rearranging, we obtain the transformed Hamiltonian as

$$H' = \frac{\beta}{\zeta_1}\sqrt{m^2 + \zeta_1^2 p^2}, \tag{4.19}$$

where we see that the effect of noncommutativity is included in the new transformed Hamiltonian, in the form of ζ_1. Of course, if $\beta(l^2) = 0$, then we have $\zeta_1 = 1$ and we obtain the known transformed Hamiltonian [16] as

$$H' = \beta\sqrt{m^2 + p^2}.$$

Incidentally, we have also found the modified unitary transformation as

$$U_L = e^{\beta\alpha \cdot \vec{p}\theta(\vec{p})} = \exp\left[\frac{1}{2}\beta\alpha \tan^{-1}(\frac{\zeta_1 p}{m})\right], \qquad (4.20)$$

where the subscript L refers to the low-energy scenario. This is the modification of the Foldy–Wouthuysen transformation [17, 18]. Thus, we see that taking into consideration the noncommutative nature of spacetime one gets new effects. Next, we shall consider Eq. (4.16) and find out if it leads to the high-energy scenario. We consider a similar type of unitary transformation as before, namely, relation (4.17). Therefore, we have the transformed Hamiltonian as

$$
\begin{aligned}
H' &= \quad (\cos p\theta + \beta\frac{\alpha \cdot \vec{p}}{p}\sin p\theta)((\alpha \cdot \vec{p} - \beta\frac{m}{\zeta_2}))(\cos p\theta - \beta\frac{\alpha \cdot \vec{p}}{p}\sin p\theta) \\
&= \quad \alpha \cdot \vec{p}(\cos 2p\theta + \beta\frac{m}{\zeta_2 p}\sin 2p\theta) + \beta(\frac{m}{\zeta_2}\cos 2p\theta - p\sin 2p\theta).
\end{aligned}
$$

Now, putting

$$\tan 2p\theta = \frac{m}{\zeta_2 p} \qquad (4.21)$$

in the second term of the last line, we obtain

$$H' = \alpha.\mathbf{p}(\cos 2p\theta + \beta\frac{m}{\zeta_2 p}\sin 2p\theta).$$

Squaring both sides, using (4.21) and rearranging the terms, we would obtain

$$H' = \frac{\alpha}{\zeta_2}\sqrt{m^2 + \zeta_2^2 p^2} \qquad (4.22)$$

which is the new transformed Hamiltonian for the high-energy scenario, considering the noncommutative effects. As usual, for $\beta(l^2) = 0$, we have $\zeta_2 = 1$ and the usual Hamiltonian

$$H' = \alpha\sqrt{m^2 + p^2}.$$

We note that both the low-energy and the high-energy scenarios are presented in the unified formulation. Thus, we have obtained the following unitary transformation:

$$U_H = e^{\beta\alpha.\mathbf{p}\theta(\mathbf{p})} = \exp[\frac{1}{2}\beta\alpha \tan^{-1}(\frac{m}{\zeta_2 p})], \qquad (4.23)$$

where the subscript H refers to the high-energy scenario and relation (4.22) is the modified Cini–Touschek transformation. We have corroborated the fact that the Klein–

Gordon formulation is the combination of both the Foldy–Wouthuysen and the Cini–
Toushek formulation, where the former refers to the low-energy case and the latter to
the high-energy case. Besides, let us take a look at the transformed Hamiltonians that
we have derived. For the low- and high-energy scenarios, we have, respectively, relations (4.19) and (4.22). Thus, from the modified Foldy–Wouthuysen transformation,
we can derive

$$H'_L = U_L H_L U_L^{-1} = \beta E'_L, \tag{4.24}$$

where E'_L denotes the modified energy levels for the low-energy scenario $(m^2 \gg p^2)$
given as

$$E'_L = \frac{1}{\zeta_1} \sqrt{m^2 + \zeta_1^2 p^2}.$$

Again, from the modified Cini–Toushek transformation we have

$$H'_H = U_H H_H U_H^{-1} = \alpha E'_H, \tag{4.25}$$

where E'_H denotes the modified energy levels for the high-energy scenario $(m^2 \ll p^2)$
given as

$$E'_H = \frac{1}{\zeta_2} \sqrt{m^2 + \zeta_2^2 p^2}.$$

The transformed Hamiltonians (4.24) and (4.25) are the same as the conventional
ones [18], except for the fact that the noncommutative feature of spacetime has been
included in them which has culminated in the modification of the energy levels.
Finally, we remark that both can go into one another if we introduce i in the exponential. This essentially means that compact space of four rotations goes into the
Minkowski space.

4.4   The Parameter ζ

Now, the modified Hamiltonians (4.19) and (4.22) give the modified energy levels.
Apparently, there is a shift from the known values of the energy levels which should
be observed in experiments. This fact can be correlated to the modified mass–energy
relation that has been studied in the author's previous work [21–23] given as

$$E^2 = m^2 + p^2 - \frac{\lambda^2 l^2 c^2}{\hbar^2} p^4. \tag{4.26}$$

Considering natural units, for a general case of (4.19) and (4.22), we can write

$$\frac{1}{\zeta} \sqrt{m^2 + \zeta^2 p^2} = \sqrt{m^2 + p^2 - \lambda^2 l^2 p^4}, \tag{4.27}$$

where $\lambda \approx -10^{-3}$ is a constant, whose value had been previously deduced in [21] and l is the Compton length ($l = \frac{\hbar}{mc}$). But here, due to relativistic considerations we take it as the de Broglie length ($l = \frac{\hbar}{p}$) of the particle. Incidentally, we had shown [21] that the constant λ is related to the electron gyromagnetic ratio and the Schwinger correction terms by the relation:

$$g = 2[1 + \frac{\alpha}{2\pi} + f(\alpha)] = 2[1 - \lambda], \tag{4.28}$$

where α is the fine structure constant and $f(\alpha)$ consists of higher orders of α. From the above equation, one can easily see that

$$\lambda \approx -\frac{\alpha}{2\pi}$$

neglecting the higher order terms of α. This result also helped in explaining the *GZK cutoff* and the *Lamb shift* phenomenon, as we saw in previous work [24]. Essentially, $|\lambda| \approx \frac{\alpha}{2\pi}$ is the reduced fine structure constant. However, squaring both sides of (4.27) and after some rearrangement we derive the following result for the parameter ζ

$$\zeta = (1 - \varepsilon^2)^{-\frac{1}{2}}, \tag{4.29}$$

where $\varepsilon = \frac{\lambda l p^2}{m}$, in terms of natural units ($\hbar = c = 1$). Particularly, we know that the de Broglie length (l) is given as

$$l = \frac{\hbar}{p}$$

or reverting to ordinary units we have

$$\varepsilon = \frac{\lambda l c}{\hbar} \frac{p^2}{mc^2}.$$

Using the previous relation for l and $p = mc$ the value of ε is

$$\varepsilon = \lambda c \frac{mc}{mc^2}$$
$$= \lambda.$$

Now, if we hadn't considered $p = mc$ then we would have obtained

$$\varepsilon = \lambda \frac{p}{mc}. \tag{4.30}$$

Thus, we would have the general value of ζ as

$$\zeta = \frac{1}{\sqrt{1 - \lambda^2 \frac{p^2}{m^2 c^2}}}. \tag{4.31}$$

Now, from Eq. (4.31) we can infer three possible cases (with $\lambda = -10^{-3}$), as follows:

1. For, $p = mc$ we would have

$$\zeta \approx 1.0000005.$$

2. For, $p \ll mc$ (non-relativistic scenario) we have ζ nearly equal to 1 but

$$\zeta > 1.$$

3. For, the case $p \gg mc$ (ultra-relativistic scenario) we must have

$$(\frac{p}{mc})_{\text{max}} < +10^3 \tag{4.32}$$

which is a critical value, in the absence of any interactions. If $(\frac{p}{mc})_{\text{max}} \geq 10^3$ then we would have either an infinite or an imaginary value of ζ which would make Equations (4.19) and (4.22) unrealistic and unphysical. Now, from special relativity we have the following relation:

$$p = \gamma mv,$$

where γ is the Lorentz factor and v is the velocity of the particle under consideration. Thus, we may write

$$(\frac{p}{mc})_{\text{max}} = (\frac{\gamma v}{c})_{\text{max}} < 1000$$

giving

$$\frac{1}{\sqrt{1 - (\frac{v}{c})^2_{max}}}(\frac{v}{c})_{max} < 1000.$$

From here, we would obtain

$$(\frac{v}{c})_{\text{max}} < 0.9999995 \tag{4.33}$$

and the corresponding Lorentz factor as

$$(\gamma)_{\text{max}} < 10^6 \tag{4.34}$$

Thus, we can infer that for a spin $-\frac{1}{2}$ particle obeying (4.15) or (4.16), at the ultra-relativistic limit, the Lorentz factor and the factor $\frac{v}{c}$, both have an upper bound, in the absence of any field or interaction. To be precise $\frac{v}{c} < 1$, with the noncommutative nature of spacetime in case of the particles referred to, there is a limit to velocities. This upper bound is borne out of several observations [25–29]. So when the noncommutative nature of spacetime is taken into consideration we have such feasible results. Now, neglecting this noncommutativity takes us back to the known scenarios and results. But the limit to the Lorentz factor is a non-trivial derivation since it might provide further insights into ultra-relativistic phenomena for particles obeying the relations (4.15) and (4.16), where the former concurs with the low-energy scenario (Foldy–Wouthuysen case) and the latter with the high-energy scenario (Cini–Toushek case). It must be borne in mind that these conclusions are for ordinary particles with an invariant mass, but not for neutrinos which are different, as seen.

4.5   The Lorentz Factor Inherent in Noncommutativity

In the light of the approach considered in this chapter, we see that the noncommutative feature of spacetime plays an important role in the understanding of several phenomena. Particularly, as we saw, this inherent noncommutativity puts a restriction on the Lorentz factor. This insight can be extended to achieve further interesting results. Let us consider the Lorentz factor and see if we can connect our approach with the acceleration of the universe. We have

$$\gamma = \frac{1}{\sqrt{1 - (\frac{v}{c})^2}}$$

from a purely heuristic point of view, just to get a feel, we differentiate both sides with respect to time we can get the acceleration as

$$a = \frac{dv}{dt} = \frac{c}{\sqrt{1 - \frac{1}{\gamma^2}}} \times \frac{1}{\gamma^3} \frac{d\gamma}{dt}. \tag{4.35}$$

Strictly speaking, there should be a factor $\mathcal{Q}$ to the time derivative, in this equation to take into account, effects like time dilatation. If we take the limiting value of the Lorentz factor as proposed above to be

$$\gamma \approx 10^6$$

and write

$$\frac{d\gamma}{dt} = \Delta$$

then from (4.35) we get the acceleration as

$$a \approx 1.0000005 \times 10^{-8} \Delta \ \mathrm{cm/s^2},$$

where Δ is small since γ is bounded. Thus, we can simply drop the "Δ" factor and finally write

$$a < 1.0000005 \times 10^{-8} \text{ cm/s}^2. \tag{4.36}$$

This is interesting as it is almost exactly of the order of the acceleration produced by the cosmological constant [30] which is given by $\frac{c^2}{R}$ (where R is the radius of the universe). In an altogether different approach, the bound in (4.36) was derived earlier by the author [31]. Also, this result corresponds to the anomalous acceleration of the Pioneer 10 and 11 [30, 32].

4.6   Minkowski Spacetime in the Light of Noncommutative Geometry of Modern Quantum Gravity Approaches

The Minkowski spacetime is a combination of three-dimensional Euclidean space and time into a four-dimensional manifold where the spacetime interval between any two events is independent of the inertial frame of reference in which they occur. In this section, we investigate properties of the Minkowski space with the additional consideration of noncommutativity. We have shown [15] that space noncommutativity also implies momentum noncommutativity. One can begin with the following relation:

$$[p_x, p_y] = \eta\theta(l^2), \tag{4.37}$$

where η generally takes the value of "$\pm i$" and $\theta(l^2)$ is a matrix with elements that are functions of the minimum fundamental length l. Considering a representation of the momentum as

$$p_i = P_i\sigma_i,$$

where σ's are the Pauli matrices, one can deduce a relation of the form:

$$v_x v_y < \frac{g(l^2)}{m^2}$$

which shows that the individual component velocities ($v_i's$) are bounded, i.e. the velocity cannot be infinite. This boundedness brings out the relativistic feature of Minkowski spacetime which prohibits infinite velocities.

4.7   Dark Energy and Spacetime Geometry

In this section, we try to find a connection between the noncommutative nature of spacetime and the *zero point energy*. We observe that extra effects come into play when we take into account the Compton scale effects in such a spacetime and the electromagnetic field tensor and the current density get modified. This defines an underlying connection between noncommutativity and the *zero point energy*.

In contrast to the usual commutative spacetime, a noncommutative spacetime has been considered and studied by many authors including the author [5] as already noted, as the framework of various fundamental phenomena. As noted, this noncommutative nature of spacetime was investigated by H. Snyder in the context of the infrared catastrophe of soft photons in the Compton scattering and in general to renormalize quantum field theory by applying the noncommutative quantized spacetime. Besides, such a framework has been widely studied by other authors [2, 3, 33–36] also. We begin with the following fundamental relation of noncommutativity [7] as

$$[x_\mu, x_\nu] = \varepsilon \beta(l^2), \tag{4.38}$$

where ε is a constant and $\beta(l^2)$, a suitable matrix, is some linear function of the square of the Compton length (l) of the electron which is very small. Here, l is the minimal physical length. Normally, in modern quantum gravity approaches, the minimum length l is taken to be the Planck length [37, 38]. But over the years the author has considered l to be the Compton wavelength [39–42]. To see how this works let us go back to the Dirac coordinate [9]

$$x = (c^2 p_1 H^{-1} t) + \frac{l}{2} c\hbar(\alpha_1 - cp_1 H^{-1}) H^{-1} \tag{4.39}$$

with similar expressions for the other two coordinates. The first term in (4.39) gives the usual position coordinates which commute with one another. It is the second term which gives the *zitterbewegung* spread over the Compton wavelength. In other words, as noted, our usual spacetime coordinates are real-valued averages over the Compton wavelength [9]. Let us analyse (4.39) in a little greater detail. Let us write it as

$$x_l = \bar{x}_l + \Theta_{lk} \bar{p}_k, \tag{4.40}$$

$$p_l = \bar{p}_l, \tag{4.41}$$

where the x's, $\bar{x}$ and $\bar{p}$ obey the usual commutation relations.

$$[\bar{x}_l, \bar{x}_k] = 0, \tag{4.42}$$

$$[\bar{x}_l, \bar{p}_k] = \imath \hbar \delta_{lk}, \tag{4.43}$$

$$[\bar{p}_l, \bar{p}_k] = 0, \tag{4.44}$$

where $\bar{x}$ represents the averaged space coordinate, that is, the first term in (4.39). However, it is easy to verify that for the x's and p's, as can be easily verified

$$[x_l, x_k] = -2\imath \hbar \Theta_{lk}, \tag{4.45}$$

$$[x_l, p_k] = \imath \hbar \delta_{lk}, \tag{4.46}$$

$$[p_l, p_k] = 0, \tag{4.47}$$

which alternatively follows from the Snyder treatment if the Compton wavelength is taken as the minimum length. It may be further pointed out that as noted by Wigner and Salecker [43] there can be no physical measurements within the Compton wavelength.

At this point we would like to emphasize the well-known close relationship between the *zero point energy*, which manifests itself even when no external fields are applied and *zitterbewegung* which, as Schrodinger first noticed—and this can be attributed to the zero point fluctuations [42, 44]. Indeed all of space is filled with the *zero point energy* (Cf. Ref. [44]) or dark energy.

In fact, as noted, this was the basis for the author's 1997 cosmology [45], which, as we saw correctly predicted that the universe would be accelerating with a small cosmological constant, at a time when the ruling paradigm was exactly the opposite.

Now, let us consider the antisymmetric field tensor in the electromagnetic case as

$$F_{\mu\nu} = -F_{\nu\mu}. \tag{4.48}$$

Now, we know that this field tensor satisfies the relation

$$\frac{\partial^2 F_{\mu\nu}}{\partial x_\mu \partial x_\nu} = 0$$

in the simple electromagnetic case. Here, the left-hand side can also be written as

$$\frac{\partial^2 F_{\mu\nu}}{\partial x_\mu \partial x_\nu} = \frac{1}{2}[\frac{\partial^2 F_{\mu\nu}}{\partial x_\mu \partial x_\nu} + \frac{\partial^2 F_{\nu\mu}}{\partial x_\mu \partial x_\nu}]$$

which gives (upon interchanging the indices μ and ν in the second term on the right-hand side)

$$\frac{\partial^2 F_{\mu\nu}}{\partial x_\mu \partial x_\nu} = \frac{1}{2}[\frac{\partial^2 F_{\mu\nu}}{\partial x_\mu \partial x_\nu} - \frac{\partial^2 F_{\mu\nu}}{\partial x_\nu \partial x_\mu}].$$

Now, from relation

$$\frac{\partial^2}{\partial x_\mu \partial x_\nu} = [1 - \frac{\varepsilon\beta(l^2)}{\delta x_\nu \delta x_\mu}]\frac{\partial^2}{\partial x_\nu \partial x_\mu}, \tag{4.49}$$

we have

$$\frac{\partial^2 F_{\mu\nu}}{\partial x_\mu \partial x_\nu} = \frac{1}{2}\left[\frac{\partial^2 F_{\mu\nu}}{\partial x_\mu \partial x_\nu} - \frac{\partial^2 F_{\mu\nu}}{\partial x_\nu \partial x_\mu}\right] = -\frac{1}{2}\left[\frac{\varepsilon\beta(l^2)}{\delta x_\nu \delta x_\mu}\right]\frac{\partial^2 F_{\mu\nu}}{\partial x_\nu \partial x_\mu}.$$

Using relation (4.48) this yields

$$\frac{\partial^2 F_{\mu\nu}}{\partial x_\mu \partial x_\nu} = \frac{1}{2}\left[\frac{\varepsilon\beta(l^2)}{\delta x_\nu \delta x_\mu}\right]\frac{\partial^2 F_{\nu\mu}}{\partial x_\nu \partial x_\mu}. \tag{4.50}$$

Here, as before, we use the Snyder noncommutative relation in the d'Alembertian, the right-hand side arises due to the vacuum fluctuations of the electromagnetic field in a noncommutative spacetime. Now, interchanging μ and ν ($\mu \leftrightarrow \nu$) on the right-hand side of the above equation and writing

$$\frac{1}{2}[\frac{\beta(l^2)}{\delta x_\nu \delta x_\mu}]\frac{\partial^2 F_{\mu\nu}}{\partial x_\mu \partial x_\nu} = \frac{\partial^2 (F_{\mu\nu})_0}{\partial x_\mu \partial x_\nu},$$

where $(F_{\mu\nu})_0$ is the field tensor for the *zero point energy*, we get

$$\frac{\partial^2 F'_{\mu\nu}}{\partial x_\mu \partial x_\nu} = 0 \tag{4.51}$$

or

$$\frac{\partial j'_\mu}{\partial x_\mu} = 0, \tag{4.52}$$

where

$$F'_{\mu\nu} = F_{\mu\nu} - \varepsilon(F_{\mu\nu})_0 \tag{4.53}$$

and

$$j'_\mu = j_\mu - \varepsilon(j_\mu)_0 \tag{4.54}$$

are, respectively, the total field tensor and the total current density, respectively. We should remember that the last two relations are comprised of the normal electromagnetic field and the *zero point energy* or the field of the quantum vacuum. Averaging over these fluctuations

$$\frac{1}{2}[\frac{\beta(l^2)}{\delta x_\nu \delta x_\mu}]\frac{\partial^2 F_{\mu\nu}}{\partial x_\mu \partial x_\nu} = \frac{\partial^2 (F_{\mu\nu})_0}{\partial x_\mu \partial x_\nu}$$

up to the Compton scale one would obtain the actual contribution from the *zero point energy*.

Now, let us again consider the relations (4.53) and (4.54). These two equations will modify the Maxwell equations of electromagnetism owing to the modified field tensor $F'_{\mu\nu}$ in

$$F'_{\mu\nu} = (F_{\mu\nu})_0 + \xi(F_{\mu\nu}). \tag{4.55}$$

On the other hand, as seen in [21] we had considered the following identification based on the presence of the *zero point energy*

$$F'_{\mu\nu} = (F_{\mu\nu})_0 + \xi(F_{\mu\nu}) \tag{4.56}$$

corresponding to a modified vector potential

$$A'_\mu = (A_\mu)_0 + \xi A_\mu$$

and a current density

$$j'_\mu = j_{\mu 0} + \xi j_\mu,$$

where $(A_\mu)_0$ is the *vector potential* associated with the *zpe*, $j_{\mu 0}$ is the current density corresponding to $(A_\mu)_0$, and ξ is a convenient constant which can be taken to be unity for the sake of simplicity. With these considerations it was shown that considering a modified field tensor $F'_{\mu\nu}$, the *anomalous* gyromagnetic ratio for the electron is explained very elegantly as we know [21] including Schwinger's correction terms. In fact, it has been shown there that the current density $j_{\mu 0}$ in the Compton scale is the reason for the anomaly in the *gyromagnetic ratio* of the electron. More precisely, if we set $\varepsilon = -1$ and $\xi = 1$ in (4.55), then we have the same equations. Thus, we are able to see a rationale for Eqs. (4.53) and (4.54) that these extra effects give rise to certain phenomena that occur due to *zitterbewegung* effects in the Compton scale. On the other hand, it is clear from Eqs. (4.53) and (4.54) that if $\varepsilon = 0$ then we get back the usual covariant Maxwell's equations due to the normal field tensor $F_{\mu\nu}$. In this manner, we find that the noncommutative nature of spacetime and the *zero point energy* is intrinsically connected to the Compton scale. It is natural that we get the extra effects since the Compton length l and consequently $\beta(l^2)$ is extremely small it is easy to conceive that the extra effects are almost negligible. Therefore, it is natural that even if we neglect $\beta(l^2)$ or set $\varepsilon = 0$ we get relevant results. But, it is undeniable that there are some extra effects when we are in the domain of the Compton scale.

Now, we have a relation between the electromagnetic tensor $(F_{\mu\nu})$ and the *zpe* tensor $(F_{\mu\nu})_0$ as

$$\frac{1}{2}\left[\frac{\beta(l^2)}{\delta x_\nu \delta x_\mu}\right]\frac{\partial^2 F_{\mu\nu}}{\partial x_\mu \partial x_\nu} = \frac{\partial^2 (F_{\mu\nu})_0}{\partial x_\mu \partial x_\nu}. \tag{4.57}$$

From this relation, we see that the total contribution of the *zero point energy* arises from averaging over the vacuum fluctuations of the electromagnetic field up to the Compton scale. The peculiar nature of this relation is due to the domain of the Compton scale, where *zitterbewegung* effects are also present. In fact, as reiterated earlier, we may state that *zitterbewegung* is due to the dark energy. Innumerable possibilities open due to the above consideration. It has been also shown by the author [6] and others [46] that the Dirac equation gets modified due to the noncommutative nature of spacetime. This modification, as noted, provides a remarkable explanation for the *Lamb shift* [21] in the energy levels of the hydrogen atom. Hence, we can perceive that the noncommutative nature of spacetime, the Compton scale, and the *zero point energy* are very significant in the sense that they can explain several phenomena that are inexplicable by conventional ideas. It appears that noncommutativity and ZPE are intimately connected.

4.8 Going Beyond the Standard Model*

In this section, we argue that we can account for the shortcomings of the standard model by including noncommutative geometry [47].[1]

It is well known that the standard model of particle physics is as of now the most complete theory and yet there are frantic efforts to go beyond the standard model to overcome its shortcomings. Some of these are

1. It fails to deliver the mass to the neutrino which thus remains a massless particle in this theory.
2. Apart from this, it does not include gravity, which is otherwise one of the four fundamental interactions.
3. There is the hierarchy problem, viz. the wide range of masses for the elementary particles or even for the quarks.
4. It appears that other as yet undiscovered particles exist which could change the picture, for example, in supersymmetry in which the particles have their super-symmetric counterparts.
5. The standard model has no place for dark matter, which, on the other hand, has not yet been definitely found. Nor is there place for dark energy.
6. Finally, one has to explain the 18 odd arbitrary constants which creep into the theory.

There are however obvious shortcomings which can be addressed in a relatively simple manner which could enable us to go beyond the standard model. Let us start with the standard model Lagrangian [48]

$$LGWS = \sum_f (\bar{\Psi}_f (\iota \gamma^\mu \partial_\mu - m_f)\Psi_f - eQ_f \Psi_f \gamma^\mu \Psi_f A_\mu) +$$

$$+ \frac{g}{\sqrt{2}} \sum_\iota (\bar{a}^\iota_L \gamma^\mu b^\iota_L W^+_\mu + \bar{b}^\iota_L \gamma^\mu a^\iota_L W^-_\mu) + \frac{g}{2C_w} \sum_f \bar{\Psi}_f \gamma^\mu (I^3_f - 2S^2_w Q_f - I^3_f \gamma_5)\Psi_f Z_\mu +$$

$$- \frac{1}{4}|\partial_\mu A_\nu - \partial_\nu A_\mu - \iota e(W^-_\mu W^+_\nu - W^+_\mu W^-_\nu)|^2 - \frac{1}{2}|\partial_\mu W^+_\nu - \partial_\nu W^+_\mu +$$

$$- \iota e(W^+_\mu + A_\nu - W^+_\nu \Lambda_\mu) \mid \iota g' c_w (W^+_\mu Z_\nu - W^+_\nu Z_\mu|^2 +$$

---

[1] *Invited talk at Frontiers of Fundamental Physics International Symposia Number 15, Orihuelle, Spain, November 2017.

$$-\frac{1}{4}|\partial_\mu Z_\nu - \partial_\nu Z_\mu + \imath g' c_w (W_\mu^- W_\nu^+ - W_\mu^+ W_\nu^-)|^2 +$$

$$-\frac{1}{2}M_\eta^2 \eta^2 - \frac{g M_\eta^2}{8 M_W}\eta^3 - \frac{g'^2 M_\eta^2}{32 M_W}\eta^4 + |M_W W_\mu^+ + \frac{g}{2}\eta W_\mu^+|^2 +$$

$$+\frac{1}{2}|\partial_\mu \eta + \imath M_Z Z_\mu + \frac{\imath g}{2 C_w}\eta Z_\mu|^2 - \sum_f \frac{g}{2}\frac{m_F}{M_W}\bar{\Psi}_f \Psi_f \eta \qquad (4.58)$$

which includes the Dirac Lagrangian among other things. We would now like to point out that all this has been on the basis of the usual point spacetime which is what may be called commutative.

We are then left with no points but minimum intervals.

All this leads to a noncommutative geometry. One model for this, which we saw, is that of Snyder [7, 49] applied at the Compton wavelength. This leads to the so-called Snyder–Sidharth dispersion relation, the geometry being given by [5] as seen

$$[x_\imath, x_j] = \beta_{\imath j} \cdot l^2. \qquad (4.59)$$

As described in detail in Ref. [50] this leads to a modification in the Dirac and also the Klein–Gordon equation. This is because (4.59) in particular leads to the following energy–momentum relation (Cf. Ref. [5]):

$$E^2 - p^2 - m^2 + \alpha l^2 p^4 = 0, \qquad (4.60)$$

where α is a scalar constant, $\sim 10^{-3}$ [21, 51]. As noted, α gives the Schwinger term.

4.9 Gamma Matrices and Bilinear Covariants

Let us introduce the well-known notation [52] γ^μ :

$$\gamma^0 = \beta$$
$$\gamma^i = \beta \alpha^i \qquad i = 1, 2, 3$$

$$\gamma^0 = \begin{pmatrix} I & 0 \\ 0 & -I \end{pmatrix} \quad \gamma^i = \begin{pmatrix} 0 & \sigma^i \\ -\sigma^i & 0 \end{pmatrix} \quad \beta = \begin{pmatrix} I & 0 \\ 0 & -I \end{pmatrix} \quad \alpha^i = \begin{pmatrix} 0 & \sigma^i \\ \sigma^i & 0 \end{pmatrix}.$$

In terms of unit matrix I and Pauli σ^i matrices. By forming the products of the γ matrices it is possible to construct 16 linearly independent 4×4 matrices which play an important part in the theory of the Dirac equation also known as bilinear covariants. These are represented as is well known as

$$\Gamma^s = 1, \quad \Gamma^V_\mu = \gamma_\mu, \quad \Gamma^T_{\mu\nu} = \sigma_{\mu\nu}, \Gamma^P = i\gamma^0\gamma^1\gamma^2\gamma^3 = \gamma_5 \equiv \gamma^5 \quad \Gamma^A_\mu = \gamma_5\gamma_\mu$$
(4.61)

the Γ^n can be easily shown to be linearly independent [16, 52]. If we work with this energy–momentum relation (4.60) and follow the usual process we get as in the usual Dirac theory

$$\left\{\gamma^\mu p_\mu - m\right\}\psi \equiv \left\{\gamma^\circ p^\circ + \Gamma\right\}\psi = 0,$$
(4.62)

where Γ is a bilinear covariant given in Eq. (4.61). We now include the extra term in the energy–momentum relation (4.60). It can be easily shown that this leads to

$$p_0^2 - \left(\Gamma\Gamma + \{\Gamma\beta + \beta\Gamma\} + \beta^2\alpha l^2 p^4\right)\psi = 0,$$
(4.63)

whence the modified Dirac equation:

$$\left\{\gamma^\circ p^\circ + \Gamma + \gamma^5\alpha l p^2\right\}\psi = 0.$$
(4.64)

The modified Dirac equation contains an extra term. The extra term gives a slight mass for the neutrino which is roughly of the correct order, viz. $10^{-8}m_e$, m_e being the mass of the electron. The behaviour too is that of the neutrino [50, 53].

To sum up the introduction of the noncommutative geometry described in (4.59) leads to a Dirac-like equation (4.64) and a Lagrangian that leads to the neutrino mass.

It must be pointed out that the modified Lagrangian differs from the usual Lagrangian in that the γ^0 matrix is now replaced by a new matrix

$$\gamma^{0'} = \gamma^0 + \gamma^0 \cdot \gamma^5 l p^2$$

that includes the term which gives rise to the neutrino mass. Further as has been discussed in detail the extra term arising out of the noncommutative geometry is the direct result of the dark energy which thus also features in the modified standard model Lagrangian. This apart, from a different angle, this argument has been shown to lead to a mass spectrum for elementary particles that includes all the elementary particles, giving the masses with about 5% or less error [5]. So now features like neutrino mass and dark energy are included.

4.10 Noncommutativity and Relativity

The concept of relativity maybe deduced from noncommutative spacetime. In the sense that noncommutativity puts a cap on velocities, whereas there is no such cap in Galilean relativity.

When dealing with special relativity we are led to a relationship between space and time. That is, there is a little of space in time and vice versa: they are woven into a single fabric called spacetime. Special relativity throws up several verifiable effects like contraction of length, time dilation, relativistic mass, a universal speed limit, etc. In this chapter, we show that one can obtain the relativistic effects of spacetime from noncommutativity. This work can be derived from Snyder's [7, 49] noncommutative geometry without using the Weyl–Moyal formalism [3, 8].

In the following section, from the relations of noncommutativity, we deduce that there is a maximum possible velocity in the universe that cannot reach infinity. This is used to argue that special relativity arises from the concept of noncommutativity.

The same concept of there being a maximal velocity is used in the subsequent sections to show length contraction, time dilation, and relativistic mass. Lastly, we discuss how the results of this chapter are connected to special relativity.

4.10.1   *Noncommutativity and Boundedness of Velocity*

It was shown earlier by the author some years ago in [54] that not only space non-commutativity but also the related momentum noncommutativity can be examined. This was again re-examined and vindicated in [15]. The concept may be elaborated as follows:

$$[p_x, p_y] = \eta\theta(l^2) \tag{4.65}$$

with η taking the value "$\pm i$". This was also examined by other authors [3, 55]. $\theta(l^2)$ is a 2×2 matrix. Here l is the fundamental minimum length which could be the Planck length or the Compton length. $\theta(l^2)$ in can be represented as

$$\theta(l^2) = f(l^2)\theta_{xy},$$

where $f(l^2)$ is a positive, finite, and real-valued scalar and θ_{xy} is a 2×2 matrix. Further, for the momenta, we have

$$p_x = P_x\sigma_x$$

and

$$p_y = P_y\sigma_y,$$

where P_x and P_y are the scalar values of the momenta and the σ's are the Pauli matrices. Thus we have from (4.65)

$$(P_x\sigma_x)(P_y\sigma_y) - (P_y\sigma_y)(P_x\sigma_x) = \eta f(l^2)\theta_{xy}$$

$$P_x P_y[\sigma_x\sigma_y - \sigma_y\sigma_x] = \eta f(l^2)\theta_{xy}$$

which gives

$$P_x P_y[\sigma_x, \sigma_y] = \eta f(l^2)\theta_{xy}. \tag{4.66}$$

The well-known commutation relations of the Pauli matrices given by

$$[\sigma_x, \sigma_y] = 2i\varepsilon_{xyz}\sigma_z$$

whence from (4.66)

$$P_a P_b[\varepsilon_{abc}\sigma_c] = -\frac{i}{2}\eta f(l^2)\theta_{ab}, \tag{4.67}$$

where a, b, c can take values x, y, z but $a \neq b \neq c$. In order to make relation (4.67) non-trivial, the Levi-Civita tensor is non-zero. Now, for even and odd permutations of a, b and c, the sign of the Levi-Civita tensor is adjusted with the sign of $\eta = \pm i$. Therefore, we have

$$(P_a P_b)\sigma_c = \frac{1}{2}f(l^2)\theta_{ab}. \tag{4.68}$$

Multiplying both sides of (4.67) by σ_c we have

$$(P_a P_b)\sigma_c^2 = f(l^2)\theta_{ab}\sigma_c.$$

Using

$$\sigma_c^2 = I$$

we have

$$(P_a P_b)I = f(l^2)\beta_{ab}, \tag{4.69}$$

where I is the identity matrix and $\beta_{ab} = \theta_{ab}\sigma_c$ is another 2×2 matrix. Now, since relation (4.69) is an equality, it holds good even after taking determinants

$$|(P_a P_b)I| = |f(l^2)\beta_{ab}|. \tag{4.70}$$

Since, on both sides, we have scalars multiplying the 2×2 matrices, we have

$$(P_a P_b)^2|I| = \{f(l^2)\}^2|\beta_{\mu\nu}|$$

which gives

$$(P_a P_b)^2 = \{f(l^2)\}^2\varepsilon^2,$$

where

$$|\beta_{ab}| = |\theta_{ab}\sigma_c| = \varepsilon^2.$$

In the matrix θ_{ab}, ε is finite and real. Since the product of the scalar values of the momenta has to be positive, we have

$$P_a P_b = \varepsilon f(l^2) \tag{4.71}$$

and the momentum–velocity relation is

$$P = mv$$

therefore we have from (4.71) that

$$v_a v_b = \frac{\varepsilon f(l^2)}{m^2}.$$

Again, $f(l^2)$ is a finite-valued function of the square of the fundamental length. It is therefore independent of the momentum and the velocity. This makes the product $v_a v_b$ bounded. This is because, one can always find a bound, namely, a function $g(l^2)$ (finite and real valued) such that

$$g(l^2) > f(l^2).$$

Therefore, we have

$$v_a v_b < \frac{\varepsilon g(l^2)}{m^2}.$$

Consequently, the individual velocities v_a and v_b will be bounded above. From this we can conclude that the velocity of a particle is bounded

$$v \leq \alpha, \tag{4.72}$$

where α is some finite velocity and it is the maximum possible velocity in the universe.

4.10.2   Composition of Velocities and Relativity

Let S and S' be two frames of reference. The frame of reference S is at rest and S' is not at rest with respect to S. Let the velocity of S' with respect to S be ω. An observer in S' measures its velocity as ω'. Therefore, the velocity of the moving body as measured by an observer in S would be given by

$$V = \omega + \omega'. \tag{4.73}$$

From the inequality (4.72) we must have the following

$$\omega \leq \alpha$$

$$\omega' \leq \alpha$$

and

$$V = \omega + \omega' \leq \alpha.$$

Let us now consider the case where the values of ω and ω' are such that

$$\omega = \omega' \approx 0.99999\alpha$$

the values of the velocities are valid since both ω and ω' are in agreement with relation (4.72). But, incidentally, we have

$$V = \omega + \omega' = 1.99998\alpha \tag{4.74}$$

which contradicts the bound in (4.72), since α is the maximum possible velocity in the universe. This is a contradiction which needs to be resolved. This shows that the classical method of addition of velocities does not hold. Actually, this problem can be resolved if we insert the factor ρ (< 1) such that

$$V = \rho(\omega + \omega') \leq \alpha. \tag{4.75}$$

This is a departure from the Galilean relativity within the Cartesian framework. Thus, it can be said that the notion of relativity originates from the factor ρ. This quantity takes care of the apparent discrepancy which arises when velocities close to α are taken. The factor ρ depends on the velocities ω, ω', and α. Obviously, in the non-relativistic limit, with $\rho = 1$ we recover Galilean relativity involving the simple addition of velocities. This brings out the connection between the noncommutativity of spacetime, Minkowski spacetime, and relativity itself.

4.10.3 Relativistic Mass

Let us revert back once more to the reference frames S and S'. Here S' moves with velocity v with respect to S. Let the x-axis of frame S' be designated as x'. We let two bodies each of mass m' move in opposite directions along x'. Let ω' and $-\omega'$ be their velocities along the x'-axis, for an observer in S'. If the bodies merge into a single mass, then the new body will be at rest according to the law of conservation of momentum, with respect to the frame S'. For an observer in the frame S, the velocities of the two bodies accordingly will be

$$\omega_1 = \rho_1(\omega' + v)$$

and

$$\omega_2 = \rho_2(-\omega' + v).$$

From (4.75), where ω_1 and ω_2 are the velocities along x-axis. Here, $\rho_1, \rho_2 < 1$. Let m_1 and m_2 be the masses of the two bodies for an observer in the S frame. Then the object that arises when the two individual masses merge will have the mass $(m_1 + m_2)$ moving with a velocity v. Thus, from the law of conservation of momentum we write

$$m_1\omega_1 + m_2\omega_2 = (m_1 + m_2)v.$$

Using the values for ω_1 and ω_2 we have

$$m_1[\rho_1(\omega' + v)] + m_2[\rho_2(-\omega' + v)] = (m_1 + m_2)v$$

whence

$$\frac{m_1}{m_2}[\rho_1(\omega' + v)] + \rho_2(-\omega' + v) = v\frac{m_1}{m_2} + v$$

and this leads to

$$\frac{m_1}{m_2} = \frac{v + (\omega' - v)\rho_2}{\rho_1(\omega' + v) - v}. \tag{4.76}$$

If we do not invoke relativity we must have that

$$m_1 = m_2$$

which, as will be seen cannot be the case. In (4.76) putting $m_1 = m_2$ we get

$$v + (\omega' - v)\rho_2 = \rho_1(\omega' + v) - v \tag{4.77}$$

for all values of ω', v, ρ_1, and ρ_2. This being a general relation, if it does not hold for a special case, it will not hold in general. Rearranging the terms of Eq. (4.77), we get

$$v[2 - (\rho_1 + \rho_2)] = \omega'(\rho_1 - \rho_2). \tag{4.78}$$

Now, since $\rho_1, \rho_2 < 1$ the left-hand side is positive. The right-hand side could be negative if $\rho_2 > \rho_1$. This cannot be so as it is a contradiction. Therefore, we have

$$m_1 \neq m_2$$

for relativistic cases. In non-relativistic cases, where $\rho_1, \rho_2 \approx 1$ we recover the relation:

$$m_1 = m_2.$$

In Eq. (4.76), we set

$$\frac{m_1}{m_2} = \zeta. \tag{4.79}$$

Therefore, if the velocity of the second body with respect to the system S is observed as zero, i.e. if we have the velocity $\omega_2 = 0$ then its mass is the rest mass given as

$$m_2 = m_0.$$

Now, writing $m_1 = m$, we have

$$m = \zeta m_0, \tag{4.80}$$

where m is the relativistic mass and ζ depends on the velocity v of the moving frame S'. The kinetic energy (T) of a moving body may be written as

$$T = \int_0^t F \, dr,$$

where F is generally the component of the force to the displacement dr. Alternatively,

$$T = \int_0^t \frac{(dmv)}{dt} dr$$

$$T = \int_0^v v d(\zeta m_0 v),$$

whence

$$T = \zeta m_0 v^2 - m_0 \int_0^v \zeta v \, dv. \tag{4.81}$$

The two terms on the right-hand side of (4.81) represent the energy and the rest mass of the body. What all this means is that we obtain a relation for mass as in (4.80) and a relation for the energy of a body in motion as in (4.81). We have therefore

$$t = \theta t_0. \tag{4.82}$$

It can be clearly seen that this gives the time dilation relation. Thus noncommutativity yields results of special relativity including Minkowski spacetime. Therefore, the noncommutativity of spacetime leads to the special relativistic distortion of spacetime. We conclude that special relativity is a result of *noncommutative Minkowski spacetime*. This result can also follow from purely quantum mechanical considerations as has been worked out by the author (see [5]). Similarly the mass energy can also be deduced. In non-relativistic cases, where $\rho_1, \rho_2 \approx 1$ we would obviously have

$$m_1 = m_2,$$

however,

$$\frac{m_1}{m_2} = \zeta. \tag{4.83}$$

Therefore, if the velocity of the second body with respect to the system S is observed to be stationary, i.e. velocity $\omega_2 = 0$ then the mass will be the rest mass

$$m_2 = m_0$$

replacing $m_1 = m$, we get

$$m = \zeta m_0, \tag{4.84}$$

where m is the relativistic mass and ζ is frame-velocity dependent. Next, the kinetic energy (T) of a moving body is given by

$$T = \int_0^t F \, dr,$$

where F is the component of the force given the displacement dr. This is

$$T = \int_0^t \frac{(dmv)}{dt} \, dr$$

$$T = \int_0^v v \, d(\zeta m_0 v)$$

whence

$$T = \zeta m_0 v^2 - m_0 \int_0^v \zeta v \, dv. \tag{4.85}$$

4.11 Special Relativity Derivations

(1) In this chapter, we have shown that beginning with the basic considerations of noncommutativity one can extract the results of special relativity. Let us consider, for example, the relation (4.75)

$$V = \rho(\omega + \omega') \leq \alpha.$$

Here, the factor ρ apparently depends on the velocities ω, ω', and α too, in order to satisfy the condition (4.72). Incidentally, taking a cue from special relativity, we write

$$\rho = (1 + \frac{\omega \omega'}{\alpha^2})^{-1}$$

then we have the known formula arising from special relativity, considering

$$\alpha = c,$$

i.e. the velocity of light. Similarly, in the relation (4.72), i.e.

$$l_0 = \theta l$$

the factor θ depends on the velocity v of the moving reference frame S' and will also depend on the limiting velocity α. We can write

$$\theta = (1 - \frac{v^2}{\alpha^2})^{-\frac{1}{2}}$$

which gives the Lorentz factor when we consider $\alpha = c$. The same factor θ emerges again in relation (4.82):

$$t = \theta t_0$$

and we have the known result for time dilation. Again, in case of relativistic mass, we had got the relation (4.78) where we argued that it is an invalid equation. This can be easily shown when we put $\rho_1 = \{1 + \frac{\omega' v}{\alpha^2}\}^{-1}$ and $\rho_2 = \{1 + \frac{(-\omega')v}{\alpha^2}\}^{-1}$ since the bodies have velocities ω' and $-\omega'$. It is obvious from these expressions of ρ_1 and ρ_2 that

$$\rho_2 > \rho_1.$$

Therefore, the relation (4.78) is invalid in relativistic cases. Now, considering these values of ρ_1 and ρ_2 we would have

$$\zeta = \frac{\rho_2}{\rho_1} = \frac{1 + \frac{\omega' v}{\alpha^2}}{1 + \frac{(-\omega')v}{\alpha^2}} = \frac{1}{\sqrt{1 - \frac{v^2}{\alpha^2}}}.$$

Again, setting $\alpha = c$, we have ζ as the Lorentz factor. Now, in Eq. (4.85), we had found

$$T = \zeta m_0 v^2 - m_0 \int_0^v \zeta v dv.$$

Thus, we have

$$T = \frac{m_0 v^2}{\sqrt{1 - \frac{v^2}{c^2}}} - m_0 \int_0^v \frac{v}{\sqrt{1 - \frac{v^2}{c^2}}} dv.$$

After integrating the second term, one has

$$T = \frac{m_0 v^2}{\sqrt{1 - \frac{v^2}{c^2}}} + m_0 c^2 [\sqrt{1 - \frac{v^2}{c^2}} - 1]$$

which finally gives

$$T = mc^2 - m_0 c^2$$

which is the desired result. Thus, it is clear that special relativity can be obtained from the noncommutative nature of spacetime.

(2) Now, the infinitesimal increment in velocity (u) can be seen in the relation:

$$v + u = \omega + u. \tag{4.86}$$

Equation (4.86) can also be looked upon as a virtual increase in velocity. This means that "u" leaves the physics of the system of reference frames unchanged in the long run. This infinitesimal velocity was introduced for heuristic purpose and it has no physical significance whatsoever.

4.12   Some Issues in Noncommutative Spacetime Geometry

(cf. also [5]).

In this section, we find a connection between the noncommutative nature of space time and the *zero point energy*. We observe that extra effects come into play when we take into account the Compton scale effects in such a spacetime and the electromagnetic field tensor and the current density get modified. This defines an underlying connection between noncommutativity and the *zero point energy*.

Noncommutativity

Much of physics in the twentieth century and earlier has been based on either Newtonian space geometry or the Minkowski geometry of special relativity. These space times are smooth manifolds. But as higher and higher energies were realized, on the one hand, and quantum gravity approaches evolved, on the other, it was realized that our very concept of spacetime of the twentieth century would have to be modified. Many of these new concepts define noncommutative spacetime geometry, with dramatically different consequences. In contrast to the usual commutative spacetime, a noncommutative spacetime has been considered and studied by many authors including the author himself [5], as the framework of various fundamental phenomena. In other words, our usual spacetime coordinates are real-valued averages over the Compton wavelength [9]. Let us analyse

$$x = (c^2 p_1 H^{-1} t) + \frac{\iota}{2} c\hbar(\alpha_1 - c p_1 H^{-1}) H^{-1} \tag{4.87}$$

in a little greater detail. Let us write it as

$$x_l = \bar{x}_l + \Theta_{lk} \bar{p}_k, \tag{4.88}$$

$$p_l = \bar{p}_l, \tag{4.89}$$

where the x's, $\bar{x}$ and $\bar{p}$ obey the usual commutation relations.

$$[\bar{x}_l, \bar{x}_k] = 0, \tag{4.90}$$

$$[\bar{x}_l, \bar{p}_k] = \imath\hbar\delta_{lk}, \tag{4.91}$$

$$[\bar{p}_l, \bar{p}_k] = 0, \tag{4.92}$$

where $\bar{x}$ represents the averaged space coordinate that is the first term in (4.87). However, it is easy to verify that the x's and p's as can be easily verified from Eqs. (4.88) and (4.89) satisfy

$$[x_l, x_k] = -2\imath\hbar\Theta_{lk}, \tag{4.93}$$

$$[x_l, p_k] = \imath\hbar\delta_{lk}, \tag{4.94}$$

$$[p_l, p_k] = 0. \tag{4.95}$$

This is precisely what was required, which alternatively follows from the Snyder treatment if the Compton wavelength is taken as the minimum length. It may be further pointed, as postulated that as noted by Wigner and Salecker [43] there can be no physical measurements within the Compton wavelength. At this point, we would like to emphasize the well-known close relationship between the *zero point energy*, which manifests itself even when no external fields are applied and *zitterbewegung* which, as Schrodinger first noticed—and this can be attributed to the zero point fluctuations [41, 44]. Indeed all of space is filled with the zero point energy (Cf. Ref. [44]) or dark energy.

4.13 2-D Crystals

In the 1990s, the author had explored some interesting properties of low-dimensional electrons—in two and one dimensions [56, 57]. It was only after the discovery of graphene by Geim and Novoselov some 10 years later that some of these properties got highlighted. What is of relevance here is that graphene provides a test bed for noncommutative spacetime, because the space of graphene is rather like a chess board with holes [58]. This is discussed in detail in Chap. 6. The author, in a series of papers in International Journal of Modern Physics E and International Journal of Theoretical Physics, has further explored these possibilities [59–62]. Some of the important properties that come out of these noncommutative crystal-like spaces include the fractional quantum Hall effect and the minimum conductivity

$$\sigma - 4\frac{e^2}{\hbar}. \tag{4.96}$$

Equation (4.96) is the expression of free current, except that it is very faint. We would like to comment that all these so-called magical properties of graphene are due to the noncommutative nature of the honeycomb-like graphene lattices, except that the exact geometrical pattern is irrelevant. What is of relevance is the lattice-like structure which gives rise to noncommutative behaviour. In other words, we can conclude that these properties would be valid for any two-dimensional crystal structure, not just graphene. It is interesting that the same properties, the decade on have been found to be true for stanene as reported in Nature [63]. In other words, the properties of two-dimensional crystals arising from the noncommutative nature of space are independent of the exact nature of the crystal.

Some Consequences of Noncommutativity

1. We would like to re-emphasize a few points; Snyder's original formulation was for a minimum length. We have considered this length to be the Compton length. Further, as far as the mass is concerned, this can, to a certain extent, and may fully be a manifestation of the noncommutative nature of spacetime. In the case of the neutrino, we have seen that the mass thrown up is not invariant.
2. Another point which can be made is that the principle considered here is valid even if we do not take the exact value of the velocity of light. All that matters is that there is a maximal velocity.
3. To look at these considerations, from another perspective, we need to take into account a modified form of special relativity in this scenario (see Eq. (4.60)). In momentum space, this is exactly the Snyder–Sidharth relation.
4. Finally, we may add that these effects and consequences of noncommutativity can also be obtained in Minkowski space using a different approach, namely, that of Alain Connes [64].

References

1. M.R. Douglas, N.A. Nekrasov, Noncommutative field theory. Rev. Mod. Phys. **73**(4), 977 (2001)
2. S. Meljanac, D. Meljanac, A. Samsarov, M. Stojić, κ-deformed Snyder spacetime. Mod. Phys. Lett. A **25**(08), 579–590 (2010)
3. M. Chaichian, M.M. Sheikh Jabbari, A. Tureanu. Hydrogen atom spectrum and the lamb shift in noncommutative qed. Phys. Rev. Lett. **86**(13), 2716 (2001)
4. L.H. Kauffman, Noncommutativity and discrete physics. Phys. D **120**(1–2), 125–138 (1998)
5. B.G. Sidharth, *The Thermodynamic Universe* (World Scientific, Singapore, 2008)
6. B.G. Sidharth, Noncommutative spacetime, mass generation, and other effects. Int. J. Mod. Phys. E **19**(01), 79–90 (2010)
7. H.S. Snyder, The electromagnetic field in quantized spacetime. Phys. Rev. **72**(1), 68 (1947)
8. R. Horvat, D. Kekez, J. Trampetić, Spacetime noncommutativity and ultrahigh energy cosmic ray experiments. Phys. Rev. D **83**(6), 065013 (2011)

9. P.A.M. Dirac. *The Principles of Quantum Mechanics*, volume 27 (Oxford University Press, 1981)
10. K. Huang, *Statistical Mechanics* (Wiley Eastern, New Delhi, 1975)
11. D. Hestenes, The zitterbewegung interpretation of quantum mechanics. Found. Phys. **20**(10), 1213–1232 (1990)
12. G.D. Kaiser, Quantized fields in complex spacetime. Ann. Phys. **173**(2), 338–354 (1987)
13. H. Feshbach, F. Villars, Elementary relativistic wave mechanics of spin 0 and spin 1/2 particles. Rev. Mod. Phys. **30**(1), 24–25 (1958)
14. J. Kouneiher, B.G. Sidharth, Mass generation without the Higgs mechanism. Int. J. Theor. Phys. **54**(9), 3044–3082 (2015)
15. B.G. Sidharth et al., Space and momentum space noncommutativity: Monopoles. Chaos, Solitons & Fractals **96**, 85–89 (2017)
16. J.D. Bjorken, S.D. Drell. *Relativistic Quantum Mechanics* (Mcgraw-Hill College, 1964)
17. L.L. Foldy, S.A. Wouthuysen, On the Dirac theory of spin 1/2 particles and its non-relativistic limit. Phys. Rev. **78**(1), 29 (1950)
18. P.Y. Pac, On the transformation properties of the Dirac equation. Prog. Theoret. Phys. **21**(4), 640–652 (1959)
19. Y.N. Obukhov, On gravitational interaction of fermions. Fortschr. Phys. **50**(5–7), 711–716 (2002)
20. M. Cini, B. Touschek, The relativistic limit of the theory of spin 1/2 particles. Il Nuovo Cimento **7**(3), 422–423 (1958)
21. B.G. Sidharth et al., The Anomalous Gyromagnetic Ratio. Int. J. Theor. Phys. **55**(2), 801–808 (2016)
22. B.G. Sidharth et al., Lorentz invariance violation: modification of the compton scattering and the GZK cutoff and other effects. Int. J. Theor. Phys. **55**(5), 2541–2547 (2016)
23. B.G. Sidharth et al., Extra relativistic effects and the Chandrasekhar limit. Gravitation Cosmol. **22**(3), 299–304 (2016)
24. B.G. Sidharth et al., Revisiting the Lamb shift. Electron. J. Theor. Phys. **12**(34), 15–34 (2015)
25. C. Sivaram, K. Arun. Possible upper limits on Lorentz Factors in High Energy Astrophysical Processes (2010). arXiv:1008.5312
26. S. Mahajan, G. Machabeli, Z. Osmanov, N. Chkheidze, Ultra high energy electrons powered by pulsar rotation. Sci. Rep. **3**(1), 1–5 (2013)
27. P. Kumar, R.A. Hernández, Ž Bošnjak, R.B. Duran, Maximum synchrotron frequency for shock-accelerated particles. Mon. Notices Roy. Astron. Soc.: Lett. **427**(1), L40–L44 (2012)
28. D. Giannios, Powerful Gev emission from a-ray-burst shock wave scattering stellar photons. Astron. Astrophys. **488**(2), 55–58 (2008)
29. X. Zhao, Z. Li, J. Bai, The bulk Lorentz factors of fermi-lat gamma ray bursts. Astrophys. J. **726**(2), 89 (2010)
30. L. Smolin, *The Trouble with Physics: The Rise of String Theory, The Fall of a Science, And What Comes Next* (Houghton Mifflin Harcourt, 2007)
31. B.G. Sidharth, Effects of varying G. Nuovo Cimento della Societa Italiana di Fisica [Sezione] B **115**, 151–154 (2000)
32. J.D. Anderson et al., Study of the anomalous acceleration of Pioneer 10 and 11. Phys. Rev. D **65**(8), 082004 (2002)
33. M.V. Battisti, S. Meljanac, Modification of Heisenberg uncertainty relations in noncommutative Snyder spacetime geometry. Phys. Rev. D **79**(6), 067505 (2009)
34. M.V. Battisti, S. Meljanac, Scalar field theory on noncommutative Snyder spacetime. Phys. Rev. D **82**(2), 024028 (2010)
35. R.V. Mendes, Quantum mechanics and noncommutative spacetime. Phys. Lett. A **210**(4), 232–240 (1996)
36. G. Amelino-Camelia, V. Astuti, Misleading inferences from discretization of empty spacetime: Snyder-noncommutativity case study. Int. J. Mod. Phys. D **24**(10), 1550073 (2015)
37. N. Seiberg, E. Witten, String theory and noncommutative geometry. J. High Energy Phys. **1999**(09), 032 (1999)

38. M. Hayakawa, Perturbative analysis on infrared aspects of noncommutative QED on r4. Phys. Lett. B **478**(1–3), 394–400 (2000)
39. B.G. Sidharth, A brief note on "extension, spin and noncommutativity". Found. Phys. Lett. **15**(5), 501–506 (2002)
40. B.G. Sidharth, *The Universe of Fluctuations* (Springer, 2005)
41. B.G. Sidharth, Fuzzy noncommutative space–time: a new paradigm for a new century, frontiers of fundamental physics (2001)
42. A.O. Barut, A.J. Bracken, Zitterbewegung and the internal geometry of the electron. Phys. Rev. D **23**(10), 2454 (1981)
43. H. Salecker, E.P. Wigner, quantum limitations of the measurement of space-time distances. Phys. Rev. **109**, 571–577 (1958). (Jan)
44. Calphys Institute. Zero Point Energy (2013). http://www.calphysics.org/zpe.html
45. B.G. Sidharth, The universe of fluctuations. Int. J. Mod. Phys. A **13**(15), 2599–2612 (1998)
46. R. Andriambololona, C. Rakotonirina, A study of the Dirac-Sidharth equation. EJTP **8**(25), 177–182 (2011)
47. Going beyond the Standard model. Fundamental Physics and Physics Education Research (Springer, Cham), pp. 17–20
48. S. Weinberg, *The Quantum Theory of Fields*, vol. 2 (Cambridge University Press, 1995)
49. H.S. Snyder, Quantized spacetime. Phys. Rev. **71**(1), 38 (1947)
50. B.G. Sidharth, The mass of the neutrinos. Int. J. Mod. Phys. E **19**(11), 2285–2292 (2010)
51. B.G. Sidharth, A.D. Roy, Are black holes compatible with a varying G? EJTP, 153 (2015)
52. C. Itzykson, J.B. Zuber, *Introduction to Quantum Field Theory* (McGraw-Hill International Book Company, 1980)
53. B.G. Sidharth, Twenty years of dark energy. New Adv. Phys. **11**(2), 20–22 (2017)
54. B.G. Sidharth, Gravitation and electromagnetism. Nuovo Cimento B Serie **116**(6), 735 (2001)
55. H.O. Girotti, Noncommutative quantum mechanics. Am. J. Phys. **72**(5), 608–612 (2004)
56. B.G. Sidharth, Low dimensional electrons, in *Solid State Physics* ed. by R. Mukhopadhyay et al., vol. 41 (1999), p. 331
57. B.G. Sidharth, Anomalous Fermions. J. Stat. Phys. **95**(3), 775–784 (1999)
58. M. Mecklenburg, B.C. Regan, Spin and the honeycomb lattice: lessons from graphene. Phys. Rev. Lett. **106**, 116803 (2011). (Mar)
59. B.G. Sidharth, A brief note on energy from graphene. Int. J. Mod. Phys. E **24**(01), 1520001 (2015)
60. B.G. Sidharth, A note on the geometry of electromagnetism. Int. J. Mod. Phys. E **23**(11), 1450076 (2014)
61. B.G. Sidharth, Graphene and high energy physics. Int. J. Mod. Phys. E **23**(05), 1450025 (2014)
62. B.G. Sidharth, A brief note on the fractional quantum hall effect. Int. J. Theor. Phys. **54**(7), 2382–2385 (2015)
63. C. Cesare et al., Physicists announce Graphene's latest cousin: Stanene. Nature **524**(7563), 18–18 (2015)
64. A. Connes, *Noncommutative Geometry* (Springer, 1994)

Chapter 5
Mystery of the Missing Dark Matter

5.1 Non-Dark Matter Approaches

Many anomalous observations may sometimes be explained without resorting to dark matter. The MOND or Milgrom's modified Newtonian dynamics is one approach which does not use dark matter. MOND is an alternative approach of dynamics, and attempts to replace Newtonian dynamics and general relativity. It aims to explain the omnipresent mass deviations in the Universe, without having to resort to dark matter which would be needed if one sticks to standard dynamics. Milgrom postulated that while Newtonian dynamics is valid up to the scale of the solar system, at the galactic scale, it needs to be modified to be applicable. Even though a modification of the distance term in the gravitation law explains the flattening of the galactic rotation curves, it does not predict the mass velocity relation. So Milgrom applied a modification to Newtonian dynamics: A test particle at a distance r from a large mass M is subject to the acceleration a given by

$$a^2/a_0 = MGr^{-2},\qquad(5.1)$$

where a_0 is an acceleration, so that Newtonian dynamics is a good approximation only for accelerations much larger than a_0. But Eq. (5.1) would be true only when $a << a_0$. Both the statements in (5.1) can be combined in the heuristic relation:

$$\mu(a/a_0)a = MGr^{-2}.\qquad(5.2)$$

In (5.2) $\mu(x) \approx 1$ when $x >> 1$, and $\mu(x) \approx x$ when $x << 1$. It must be re-emphasized that (5.1) or (5.2) do not result from any known theory, but have been prescribed in an ad hoc manner to account for observations. It must be mentioned furthermore that the exact form of μ has no bearing on most of the implications of modified Newtonian dynamics or MOND. With this fix, the problem of galactic velocities is now solved.

© The Author(s), under exclusive license to Springer Nature Singapore Pte Ltd. 2025
B. G. Sidharth, *The Dark Energy Paradigm*,
https://doi.org/10.1007/978-981-96-3745-4_5

The author has a method in which the gravitational constant is varying to explain certain inexplicable phenomena.

The author's varying G method is as follows. Based on fluctuational cosmology we have

$$G = \frac{G_0}{T},\tag{5.3}$$

where G is the Newtonian constant of gravitation and T is the age of the universe. Replacing G with varying G, for ordinary Keplerian velocity of rotation of stars and galaxies, we get about 300 kms per second. As is required and this has been achieved without invoking dark matter. Surprisingly all the other standard observations like the precession of the perihelion of the planet Mercury, the shortening of the time period of binary pulsars and so on also follow from using a varying G.

But then, dark energy which has been a topic for study and investigation for a century has not been taken seriously. At the beginning of the twentieth century, Nernst, the father of the third law of thermodynamics, appears to have introduced this idea. To explain the superfluidity of Helium 4, Nernst resorted to dark energy. He hypothesized that dark energy was ubiquitous. Zeldovich and other astronomers some decades later reinvoked the concept of dark energy. But the cosmological constant problem posed a hurdle. That is, the cosmological constant would be of such a high value that the universe would be blown out of existence almost as soon as it was created (as mentioned earlier).

In 1997, the author worked on his concept of dark energy. His view was that the zero point energy should be used at the scale of the Compton wavelength $\sim \mathcal{O}(10^{-13}\text{ cm})$, rather than the Planck scale $\sim \mathcal{O}(10^{-33}\text{ cm})$. This was the key to solving the cosmological constant problem, namely, doing this leads to a small cosmological constant as required. It must be borne in mind that at that time the standard big bang model of cosmology focussed on the presence of dark matter which would slow down the expansion of the universe progressively and would bring it to a halt. In this dark matter scheme of things, the universe would be accelerating at a slow rate. But, on the contrary, this cosmic acceleration, which was observed by Perlmutter, Schmidt, and Reiss independently in 1998 and won them the Nobel Prize in 2011, was not as slow as the dark matter theory implied [1].

5.2 How Essential Is Dark Matter?

For the sake of completeness, we play the devil's advocate and consider two models which argue that dark matter is not an essential element, even though popular models postulate that it comprises roughly a fourth of a universe.

Our starting point is the relation [2, 3]

$$G = G_0\left(1 - \frac{t}{t_0}\right),\tag{5.4}$$

where G_0 is the present value of G, t_0 is the present age of the universe, and t is the time elapsed from the present epoch. Similarly one could deduce that (Cf. Ref. [3])

$$r = r_0 \left(\frac{t_0}{t_0 + t} \right). \tag{5.5}$$

In this scheme, the gravitational constant G varies slowly with time. This is suggested by the author's 1997 cosmology, which correctly predicted a dark-energy-driven accelerating universe at a time when the accepted paradigm was the standard big bang cosmology in which the universe would decelerate under the influence of dark matter (as mentioned before).

We reiterate the following: In the problem of galactic rotational curves (cf. Ref. [3]), we would expect on the basis of straightforward dynamics that the rotational velocities at the edges of galaxies would fall off according to

$$v^2 \approx \frac{GM}{r}. \tag{5.6}$$

However, it is found that the velocities tend to a constant value:

$$v \sim 300 \, \text{km/s}. \tag{5.7}$$

This as known had lead to the postulation of as yet undetected additional matter, the so-called dark matter. We observe that from (5.5) it can be easily deduced that [4]

$$a \equiv (\ddot{r}_o - \ddot{r}) \approx \frac{1}{t_o}(t\ddot{r}_o + 2\dot{r}_o) \approx -2\frac{r_o}{t_o^2} \tag{5.8}$$

as we are considering infinitesimal intervals t and nearly circular orbits. Equation (5.8) shows (Cf. Ref [5] also) that there is an anomalous inward acceleration, as if there is an extra attractive force or an additional central mass.

So,

$$\frac{GMm}{r^2} + \frac{2mr}{t_o^2} \approx \frac{mv^2}{r} \tag{5.9}$$

From (5.9) it follows that

$$v \sim \left(\frac{2r^2}{t_o^2} + \frac{GM}{r} \right)^{1/2}. \tag{5.10}$$

From (5.10), it is easily seen that at distances within the edge of a typical galaxy, that is, $r < 10^{23}$ cms the Eq. (5.6) holds but as we reach the edge and beyond, that is, for $r \geq 10^{24}$ cms we have $v \sim 10^7$ cms per second, in agreement with (5.7).

Thus, the time variation of G explains observation without invoking dark matter. It may also be mentioned that other effects like the Pioneer anomaly and shortening of the period of binary pulsars can be deduced [6], while new effects also are predicted.

Milgrom [7], as mentioned in Chap. 4, approached the problem by modifying Newtonian dynamics at large distances. This approach is purely phenomenological and ad hoc. The idea was that perhaps standard Newtonian dynamics works at the scale of the solar system but at galactic scales involving much larger distances perhaps the situation is different. However, a simple modification of the distance dependence in the gravitation law, as pointed by Milgrom would not do, even if it produced the asymptotically flat rotation curves of galaxies. Such a law would predict the wrong form of the mass velocity relation. So Milgrom suggested the following modification to Newtonian dynamics: A test particle at a distance r from a large mass M is subject to the acceleration a given by

$$a^2/a_0 = MGr^{-2}, \tag{5.11}$$

where a_0 is an acceleration such that standard Newtonian dynamics is a good approximation only for accelerations much larger than a_0. The above equation however would be true when a is much less than a_0. Both the statements can be combined in the heuristic relation:

$$\mu(a/a_0)a = MGr^{-2}. \tag{5.12}$$

In (5.12) $\mu(x) \approx 1$ when $x >> 1$, and $\mu(x) \approx x$ when $x << 1$. As noted earlier, interestingly, it must be mentioned that most of the implications of MOND do not depend strongly on the exact form of μ.

It can then be shown that the problem of galactic velocities is solved [8–12].

It is interesting to note that there is a relationship between the varying G approach, which has a theoretical base and the purely phenomenological MOND approach. Let us write

$$\beta \frac{GM}{r} = \frac{r^2}{t_0^2} \text{ or } \beta = \frac{r^3}{GMt_0^2}$$

whence

$$\alpha_0 = v^2/r = \frac{GM}{r^2} \quad \alpha = \frac{r}{t_0^2}$$

so that

$$\frac{\alpha}{\alpha_0} = \frac{r^3}{GMt_0^2} = \beta.$$

At this stage, we can see a similarity with MOND. For if $\beta << 1$ we are with the usual Newtonian dynamics and if $\beta > 1$ then we get back to the varying G case exactly as with MOND.

5.3　Dark Matter Anomalies

We argue that the most recent anomalous data on dark matter thrown up by the Planck satellite and the dark energy survey can be explained by varying G theory.

A recent article published in Nature raises serious questions about dark matter. A team lead by Reinhard Genzel of the Max Planck Institute for Experimental Physics using the 8.2 m very large telescope in Chile made a surprising finding: Some galaxies which are ten billion years old, born when the universe was just one-fifth of its current age did not display the flattening rotation curves which are a hallmark of dark matter, as is well known. Rather they displayed the normal rotation curves which can be attributed to Keplarian orbits. The conclusion is that dark matter developed in galaxies as the universe aged.

Further, even more recently, thanks to the dark energy survey (DES) and the Planck satellite more confusion has been created. The latter has shown that the dark matter was 34% when the universe was young while DES shows it has 26%. The question is, what has happened to the missing dark matter? It may also be noted that to this day neither has dark matter been observed nor its identity established.

5.4　Dark Matter: Confusion and Mystery

We now consider various observations in the past few years which give contradictory results as far as the presence of dark matter is concerned. It is well known that dark matter was hypothesized more than 70 years ago by Fritz Zwicky. The reason behind the hypothesis was that in the velocity rotation curves of galaxies in the outermost curves the velocities seem to tend to a constant value of, for example, $\sim$300 km per second, whereas in the normal case they would be tending to zero according to the Keplarian law:

$$v \propto 1/\sqrt{r}.$$

This indicated that there was some unseen or dark matter in the galaxy which was causing the deviation. Yet the fact remains that after more than eight decades we still do not have the faintest clue as to what exactly this mysterious dark matter is. Among the various speculative hypothesis advanced are: There could be some brown dwarf stars which have not been accounted for because of their low luminosity. Or even, there could be a massive black hole at the centre of the galaxies which we are not being able to access. Other hypothesis have included exotic particles, for example, sterile neutrinos, supersymmetric particles which partner ordinary particles, and so on. All these have eluded detection to date. Another hypothesis is: recently the LHC detected the heavy pentaquark. Could these pentaquarks be responsible?

Further recent observations have revealed another anomalous feature: A study by the Max Planck Institute of Experimental Physics with the 8.2 m telescope of older galaxies more than 10 billion years old reveals that there was no dark matter [13]. In this case, we have to suppose that dark matter developed at a later stage.

Further there is a mismatch between the dark energy survey and the Planck satellite [14]. While the former gives a 26% as the dark matter the latter gives 34%, prompting the question, is dark matter decreasing?

The mystery deepens further because of observations of Adam Riess who shared the Nobel Prize in 2011 along with Saul Perlmutter and Brian Schmidt. These latest observations first reported in 2015/16 indicated that the universe driven by dark energy is accelerating some 8% faster than what current cosmological models suggest. Riess reconfirmed this anomalous finding in 2017/18.

It must be noted that according to the current cosmological model, the universe is made up roughly of 23% dark matter, whatever that is and about a little over 90% of dark energy, with ordinary matter constituting roughly 4–5%. So from where does the extra acceleration come?

The author and others [15] have calculated that if there were no dark matter at all or much less of it, then there would appear the extra 8% acceleration we are looking for. Indeed a few physicists and astronomers now suspect that dark matter could be just ordinary matter in "disguise".

In fact the author's work on a time-varying gravitational constant could provide another alternative way out. In this case, the gravitational constant G is given by (after the period of nucleo-synthesis there being no dark matter in the early period (Cf. Ref. [13]), [2, 3, 6, 16]:

$$G = \frac{G_0}{T},$$

where T is the age of the universe. This leads to precisely the non-Keplarian behaviour at edges of galaxies, viz. $v \sim 300\,\mathrm{m/s}$ which we actually observe today suggesting that dark matter could be an artefact of a time-varying G (Cf. also [17]).

In fact in this case we have (Cf. Ref. [2]).

$$v \approx \left(\frac{2r^2}{t_o^2} + \frac{GM}{r} \right)^{1/2}. \tag{5.13}$$

From (5.13), it is easily seen that at distances within the edge of a typical galaxy, that is, $r < 10^{23}$ cms the equation, as noted

$$v^2 \approx \frac{GM}{r} \tag{5.14}$$

holds but as we reach the edge and beyond, that is, for $r \geq 10^{24}$ cms we have $v \sim 10^7$ cms per second, in agreement with

$$v \sim 300\,\mathrm{km/s}. \tag{5.15}$$

In fact as can be seen from (5.13), the first term in the square root has an extra contribution (due to the varying G) which is roughly some three to four times the second term, as if there is an extra mass, roughly that much more.

Yet another way out would be the MOND or modified Newtonian dynamics hypothesis which has not found much favour because of its ad hoc character.

In any case in 1997, the existing standard big bang model was that of a predominantly dark matter universe with a hint of ordinary matter which was decelerating due to the dark matter gravitation. At that time, the author had proposed a model of an accelerating universe, the accelerating driven by what is now called dark energy. It was the very next year that dark energy and cosmic acceleration was observed by Perlmutter, Riess, and Schmidt. Are we today at a similar juncture?

5.5 Anomaly of Dark Matter

In this section, we see that the existence of dark matter is itself inconsistent with Hubble's law. Considering the new observational values by Riess et al. regarding the Hubble constant we have arrived at this result. Also, we attempt to study the inconsistencies in galactic rotation curves by an alternative method.

Hubble's law is regarded as one of the major observational basis for the expansion of the universe. The existence of dark matter was hypothesized by Zwicky, as noted, [18, 19] who inferred the existence of unseen matter based on his observations of the rotational velocity curves at the edge of galaxies. Although Jacobus Kapteyn [20] and Jan Oort had [21] also had come to the same conclusions before Zwicky. Since then, various efforts have been made to prove the existence of dark matter [cf. Ref. [22] for detailed review]. Recently, some experiments were conducted to detect weakly interacting massive particles (WIMPs) that interact only through gravity and the weak force. These were hypothesized as the constituents of dark matter. Such attempts failed [23].

Interestingly, authors such as Milgrom [8], Bekenstein [24] as well as the author and Mannheim [25] have tried to find alternatives to the widely accepted dark matter. The author [5, 6, 26] has also given a suitable alternative to the conventional dark matter paradigm. Nevertheless, the objective of this chapter is to assert that the existence of dark matter is inconsistent with the recent observations made by Riess [27] regarding the Hubble's constant.

The generally accepted ideas may have to be revisited in view of latest observations of Riess et al., which point to the fact that cosmic acceleration is some $5-8\%$ greater than what the usual cosmological model suggests.

5.6   The Expanding Universe

It is known that the Friedman equations govern the expansion of space in homogeneous and isotropic models of the universe within the context of general relativity. Let us begin with the following equation:

$$\mathcal{H}^2 = \left(\frac{\dot{a}}{a}\right)^2 = \frac{8\pi G}{3}\rho - \frac{kc^2}{a^2} + \frac{\Lambda c^2}{3},$$

where $\mathcal{H}$ is the Hubble parameter, a is the scale factor, G is the gravitational constant, k is the normalized spatial curvature of the universe, and Λ is the cosmological constant. Considering $k = 0$ (a flat universe) with matter and dark energy dominating, one can derive the Hubble parameter as

$$\mathcal{H}(z) = \mathcal{H}_0[\Omega_M(1+z)^3 + \Omega_{DE}(1+z)^{3(1+w)}]^{\frac{1}{2}}, \tag{5.16}$$

where z is the redshift value or the recessional velocity and the dimensionless parameter w is given by

$$P = w\rho c^2$$

P being the pressure and ρ being the density. Now, we would like to expand the function $\mathcal{H}(z)$ using the Taylor expansion about some point z_0. This yields

$$\mathcal{H}(z) = \mathcal{H}(z_0) + \frac{\mathcal{H}'(z_0)}{1!} + \cdots$$

Neglecting terms consisting higher order derivatives of the Hubble parameter and considering that $\mathcal{H}(z_0) = \mathcal{H}_0$ we have, using (5.16)

$$\mathcal{H}(z) = \mathcal{H}_0 + \frac{\mathcal{H}_0}{2}\frac{3\Omega_M(1+z_0)^2 + 3(1+w)\Omega_{DE}(1+z_0)^{3(1+w)-1}}{[\Omega_M(1+z)^3 + \Omega_{DE}(1+z)^{3(1+w)}]^{\frac{1}{2}}}(z-z_0)^2. \tag{5.17}$$

Now, we know that if dark energy originates from a cosmological constant then

$$w = -1.$$

Therefore, in such a case, we have

$$\mathcal{H}(z) = \mathcal{H}_0 + \frac{3\mathcal{H}_0}{2}\frac{\Omega_M(1+z_0)^2}{[\Omega_M(1+z_0)^3 + \Omega_{DE}]^{\frac{1}{2}}}(z-z_0)^2. \tag{5.18}$$

Now, since numerical values suggest that $\Omega_{DE} > \Omega_M$ we can use another series expansion for the denominator of the second term above to get

$$\mathcal{H}(z) = \mathcal{H}_0 + \frac{3\mathcal{H}}{2}\frac{1}{\sqrt{\Omega_{DE}}}[\Omega_M(1+z_0)^2][1 - \frac{\Omega_M(1+z_0)^2}{2\Omega_{DE}}](z - z_0)^2.$$

Thus, we can write finally

$$\mathcal{H}(z) = \mathcal{H}_0[1 + \frac{3}{2}\frac{1}{\sqrt{\Omega_{DE}}}\{\Omega_M(1+z_0)^2\}\{1 - \frac{\Omega_M(1+z_0)^2}{2\Omega_{DE}}\}](z - z_0)^2. \quad (5.19)$$

Now, considering this equation at the point $z_0 = 0$ and for $z = 1$ to give

$$\mathcal{H} = \mathcal{H}_0[1 + \frac{3}{2}\frac{\Omega_M}{\sqrt{\Omega_{DE}}}\{1 - \frac{\Omega_M}{2\Omega_{DE}}\}]. \quad (5.20)$$

Now, the standard cosmological model suggests that the universe is made up of baryonic matter, dark matter, dark energy, and some other constituents. In a nutshell, we have [28]

$$\Omega_{Baryonic} \approx 0.04$$

$$\Omega_{Darkmatter} \approx 0.23$$

$$\Omega_{Darkenergy} \approx 0.73$$

and

$$\Omega_M = \Omega_{Baryonic} + \Omega_{Darkmatter}.$$

Using all these values in (5.20) we have the Hubble parameter

$$\mathcal{H} = \mathcal{H}_0 + 0.39\mathcal{H}_0,$$

i.e. the acceleration of the universe should be approximately 39% greater than its value. But, due to recent observations, it has been shown that the acceleration is about 5–8% greater than its value. So, in fact, we should have

$$\mathcal{H} = \mathcal{H}_0 + 0.08\mathcal{H}_0.$$

If this is the case then back working, we arrive at a quadratic equation in Ω_M as

$$(1 - 0.685\Omega_M)\Omega_M - 0.045, \quad (5.21)$$

Solving this equation we have the following two values for Ω_M:

$$\Omega_M \approx 1.41 \; or \; 0.044.$$

Now, it is a fact that $\Omega_M < 1$ since the value is unphysical and therefore we have the value of Ω_M as

$$\Omega_M \approx 0.044. \tag{5.22}$$

But, this is very nearly equal to the value of Baryonic matter, i.e. $\Omega_{Baryonic}$. This suggests that

$$\Omega_{Darkmatter} \approx 0. \tag{5.23}$$

In other words, the existence of dark matter is itself inconsistent according to the latest observations of Riess et al. In such a case, the total density of the universe is given by

$$\Omega = \Omega_{Baryonic} + \Omega_{Darkenergy} \approx 0.77 \tag{5.24}$$

which is less than the critical density. This suggests that the universe will be expanding in an accelerated manner.

5.7   Alternative to the Dark Matter Paradigm

Very recently the LUX detector in South Dakota concluded [23] that it has not found any traces of dark matter. So far this has been the most precise detector. It will be recalled that dark matter was introduced in the 1930s by F. Zwicky to explain the flattening of the galactic rotational curves: With Newtonian gravity the speeds of these galactic curves at the edges should tend to zero according to the Keplerian law, $v \propto 1/\sqrt{r}$. Here r is the distance to the edge from the galactic centre. However, velocity v remains more or less constant. Zwicky explained this by saying that there is a lot more of unseen matter concealed in the galaxies, causing this discrepancy. The fact is that even after nearly 90 years, dark matter has not been detected.

The modified Newtonian dynamics approach of Milgrom [7–12] was an interesting alternative to the dark matter paradigm. The objection of this fix has been that it is too ad hoc, without any underlying theory.

The author himself has been arguing over the years [5, 6, 26] (Cf. Ref. [2] for a summary) that the gravitational constant G is not fixed but varies slowly with time in a specific way. In fact, this variation of the gravitational constant had been postulated by Dirac, Hoyle, and others from a different point of view (Cf. Refs. [2, 3]) which, for various reasons, including inconsistencies, have, in the author's scheme, exactly accounted for the galactic rotation anomaly without needing dark matter and without contradictions.

References

1. S. Perlmutter, Discovery of a supernova explosion at half the age of the Universe. Nature **391**(6662), 51–54 (1998)
2. B.G. Sidharth, *The Thermodynamic Universe* (World Scientific, Singapore, 2008)
3. J.V. Narlikar, *An Introduction to Cosmology* (Cambridge University Press, 1993)
4. B.G. Sidharth, From the neutrino to the edge of the Universe. Chaos, Solitons & Fractals **12**(6), 1101–1109 (2001)
5. B.G. Sidharth, Effects of varying G. Nuovo Cimento della Societa Italiana di Fisica [Sezione] B **115**, 151–154 (2000)
6. B.G. Sidharth, Tests for varying G. Found. Phys. Lett. **19**(6), 611–617 (2006)
7. M. Milgrom, MOND–a pedagogical review. Acta Phys. Pol., B **32**(11), 3613 (2001)
8. M. Milgrom, A modification of the Newtonian dynamics-Implications for galaxies. Astrophys. J. **270**, 371–389 (1983)
9. M. Milgrom, Solutions for the modified Newtonian dynamics field equation. Astrophys. J. **302**, 617–625 (1986)
10. M. Milgrom, On stability of galactic disks in the modified dynamics and the distribution of their mean surface brightness. Astrophys. J. **338**, 121–127, 02 (1989)
11. M. Milgrom, Dynamics with a nonstandard inertia-acceleration relation: an alternative to dark matter in galactic systems. Ann. Phys.- ANN PHYS N Y **229**, :384–415, 02 (1994)
12. M. Milgrom, Nonlinear conformally invariant generalization of the Poisson equation to 2 dimensions. Phys. Rev. E **56**, 1148–1159 (1997). (Jul)
13. R. Genzel, Förster N. M. Schreiber, H. Übler, P. Lang, T. Naab, R. Bender, L.J. Tacconi, E. Wisnioski, S. Wuyts, T. Alexander, et al., Strongly baryon-dominated disk galaxies at the peak of galaxy formation ten billion years ago. Nature **543**(7645), 397–401 (2017)
14. S. Perrenod, Planck satellite. https://darkmatterdarkenergy.com/tag/planck-satellite
15. B.G. Sidharth, Anomaly of dark matter. Mod. Phys. Lett. A **34**(19), 1950143 (2019)
16. B.G. Sidharth, Planck oscillators and gravitation. Int. J. Theor. Phys. **48**, 3421–3425, 12 (2009)
17. J.P. Uzan, Varying constants, gravitation and cosmology. Living Rev. Relativ. **14**(1), 1–155 (2011)
18. F. Zwicky, Die rotverschiebung von extragalaktischen nebeln. Helv. Phys. Acta **6**, 110–127 (1933)
19. F. Zwicky, On the masses of nebulae and of clusters of nebulae. Astrophys. J **86**, 217 (1937)
20. J.C. Kapteyn, First attempt at a theory of the arrangement and motion of the sidereal system. ApJ **55**, 542–549 (1922)
21. J.H. Oort, The force exerted by the stellar system in the direction perpendicular to the galactic plane and some related problems. Bull. Astron. Inst. Netherlands **6**, 249 (1932)
22. G. Bertone, D. Hooper, J. Silk, Particle dark matter: evidence, candidates and constraints. Phys. Rep. **405**(5–6), 279–390 (2005)
23. D.S. Akerlb, H.M. Araújo et al., Improved limits on scattering of weakly interacting massive particles from reanalysis of 2013 LUX data. Phys Rev. Lett. **116**(16), 161301 (2016)

24. J.D. Bekenstein, Relativistic gravitation theory for the modified Newtonian dynamics paradigm. Phys. Rev. D **70**(8), 083509 (2004)
25. P.D. Mannheim, Alternatives to dark matter and Dark Energy. Prog. Part. Nucl. Phys. **56**(2), 340–445 (2006)
26. B.G. Sidharth, The puzzle of gravitation. Int. J. Mod. Phys. A **21**(31), 6315–6321 (2006)
27. A.G. Riess et al., A 2.4% determination of the local value of the Hubble constant. Astrophys. J. **826**(1), 56 (2016)
28. R.A. Knop, G. Aldering et al., New constraints on ω m, ω λ, and w from an independent set of 11 high-redshift supernovae observed with the Hubble Space Telescope. Astrophys. J. **598**(1), 102 (2003)

Chapter 6
Low-Dimensional Structures

6.1 Quantum Mechanical Black Holes: Low-Dimensional Structures

According to a recent model, [4–7] the most elementary fermion, the electron can be treated as a Kerr–Newman-type black hole bounded by the Compton wavelength, which may be called a quantum mechanical black hole (QMBH). There is a naked singularity, that is, the radius becomes complex, but this is explained by the fact that inside the Compton wavelength there are negative energies manifesting themselves in the form of *zitterbewegung*. Indeed the position in the quantum mechanical case also becomes complex or equivalently the position operator is non-Hermitian. The well-known explanation for this is [8] that strictly speaking spacetime points are meaningless, while it is only spacetime intervals which are meaningful. In quantum mechanics as is well known on averaging over such intervals, the non-Hermitian position operator with complex eigenvalues goes over to a Hermitian operator with real eigenvalues. In any case, as is well known, the Kerr–Newman metric describes the field of an electron including, and this is remarkable, the quantum mechanical anomalous gyro magnetic ratio $g = 2$. Such a model explains several interesting features like the discreteness of the charge, the left-handedness of the neutrino, and so on. It also leads to a cosmology consistent with observation including a theoretical deduction of the mass, radius, and age of the universe and other cosmological parameters and the supposedly mysterious large number coincidences (cf. Refs. [2, 9]). Interestingly, the cosmological model predicts an ever-expanding and accelerating universe, which has been recently observationally confirmed ([10, 11]). On the other hand, the model also gives a rationale for weak interactions and for the structure of particles like baryons and mesons (cf. [12]). We now see how from the above model one can argue that at low temperatures fermions exhibit an anomalous character, indeed as seen in the superfluidity of He_3 [13]. We will also briefly comment on fermionic behaviour in two and one (spatial) dimensions. These considerations are corroborated by conventional theory.

In 1995, the author had put forward two papers [1, 2]. The first examined the motion of a fermion in one dimension and the other in two dimensions. To sum up, they describe motion in nanotubes and 2D materials. While nanotubes were discovered a few years thereafter, materials like graphene were discovered 10 years later.

6.2 One- and Two-Dimensional Behaviour

In connection with quantum mechanical Kerr–Newman metric considerations, we examine the two- and one-dimensional cases seen in Chap. 3. These are idealized and extreme scenarios, since, following Wheeler, spin $\frac{1}{2}$ is responsible for three dimensions [14]. Let us digress from this issue briefly, remembering also that we are dealing with constrained quantum systems. In the high-energy relativistic domain, the reduction of even a single dimension leads the considerations to be in less than the Compton wavelength. Whence the idea of inertial mass of particles and their other behaviours become dubious (cf. [4, 5]). In this scenario, we encounter for the most part, energy components which are negative and hence exhibit left-handed behaviour. Going back to the neutrino a similar behaviour is encountered. Although this is because a fermion without mass has nearly infinite or, in practice, very large Compton wavelength. This places us in the negative energy component region (see [5] for further details). This puts us in a position to follow on the lines of quantum field theory. Now, the Hamiltonian from quantum field theory is [15]

$$H = \sum_{\pm s} \int d^3 p \, E_p [b^+(p, s)b(p, s) - d(p, s)d^+(p, s)]. \tag{6.1}$$

The coefficients represent creation and annihilation operators while bb^+ and d^+d represent particle number operators with eigenvalues 1 or 0 only. Transitioning to empty states of the Dirac sea of opposite sign of energy is forbidden as these states are inaccessible. And hence the use of commutators as against anti-commutators.

These can be verified easily. The relativistic covariant equations in two and one dimensions consist of two components displaying handedness [16]. To construct a Lagrangian with invariant mass, four components are needed and we return to three-dimensional space. To show consistency with the quantum mechanical Kerr–Newman metric model, consider the Lorentz covariant equation in one (spatial) dimension, in a familiar notation [15]:

$$\left(i\gamma^\mu \partial_\mu - \frac{mc}{\hbar} \right) \psi = 0.$$

The left-handed solution is obtained, and it must be observed that the mass term (or the energy operator term) does not have the usual factor

$$\begin{pmatrix} 1 & 0 \\ 0 & -1 \end{pmatrix}$$

which is responsible for the positive and negative energy solutions and the *zitterbewegung* (cf. [8, 15]). This leads to equations like the Fermi gas equation [17]:

$$\varepsilon_F = p_F^2/2m = \left(\frac{\hbar^2}{2m}\right)\left(\frac{6\pi^2}{v}\right)^{2/3}. \tag{6.2}$$

Equation (6.2) holds for a collection of fermions. For phonons, on the other hand, the maximum frequency is given by [17]

$$\omega_m = c\left(\frac{6\pi^2}{v}\right)^{1/3} \tag{6.3}$$

or to the quantum mechanical Kerr–Newman metric which is bounded by the Compton wavelength, and inertial mass [7]. The expectation value of the velocity operator $c\vec{\alpha}$, when considering solutions of a single sign, is given by [15]

$$J^+ = \langle c\vec{\alpha}\rangle_+ = \langle c^2 p/E\rangle_+ = \langle v_{gp}\rangle,$$

(where v_{gp} is the group velocity). This is a contradiction, because $c\,\vec{\alpha}$ has eigenvalues $\pm c$, whereas we require $v_{gp} < c$, for a massive particle. In other words, both positive and negative energy solutions have to be considered or it is as if the particle has no invariant mass. Here, we are dealing with particles similar to neutrinos. So for low dimensions there is fermion–boson transmutation and also some other statistics like anyone statistics [18, 19]. It can be shown that the behaviour of the collection of fermions is as if it is below the Fermi temperature: the average energy per unit length in one dimension is given by

$$e = \frac{\pi(kT)^2}{6\hbar v_F}, \tag{6.4}$$

where $v_F \equiv \frac{\hbar\pi\frac{N}{L}}{m}$, L being the length of the one-dimensional wire and N the number of fermions therein which is the one-dimensional form of the Stephan–Boltzmann law for radiation [20]. Denoting the average interparticle distance, $\frac{L}{N} \equiv v^{1/3}$ and using [17]

$$kT_F = \left(\frac{\hbar^2}{2m}\right)\left(\frac{6\pi^2}{v}\right)^{2/3}$$

(T_F is the Fermi temperature) and

$$kT - ev^{1/3}$$

we obtain using

$$\varepsilon_m = \frac{p_m^2}{2m} = \frac{\hbar^2}{2m}\left(\frac{6\pi^2}{v}\right)^{2/3} \tag{6.5}$$

$$T = \frac{3}{5}T_F.$$

That is, the temperature is below the Fermi temperature, and also the gas is in the ground state, [17] irrespective of the temperature.

6.3 Fermions and Bosons

It is well known that fermions and bosons have different statistical behaviour obeying, respectively, Fermi–Dirac and Bose–Einstein statistics. This holds good usually, but at very low temperatures or in low dimensions this compartmentalization becomes fuzzy because of what maybe called bosonization or semionic behaviour. The author's quantum mechanical Kerr–Newman metric model in fact foretells such a bosonization effect with respect to fermions. Of course this happens at energy scales that correspond to space scales much larger than the Compton wavelength. The superfluidity of He_3 brings out this feature. Though this is usually explained in terms of the conventional Bardeen–Cooper–Schrieffer theory that describes superconductivity (BCS). This not withstanding, there are some unexplained features [13]. The author's model [38] also anticipates handedness and the blurring of Fermi–Dirac statistics in low dimensions. All this has been confirmed by recent observations and are well known. Lastly, recent observations using carbon nanotubes [21–24] have thrown up the one-dimensional conductivity of such carbon nanotubes. Also that their behaviour is like low-temperature quantum wires. This confirms the above considerations.

6.4 Anomalous Behaviour of Bosons and Fermions

The author had pointed out, back in 1995, that electrons will display strange neutrino-like properties in two and one dimensions [2, 3, 25]. In fact, a two-component equation is obeyed [26]. A form of the Dirac equation is

$$\left(\sigma^\mu \partial_\mu - \frac{mc}{\hbar}\right)\psi = 0, \tag{6.6}$$

where σ^μ denote 2×2 matrices. When the mass equals zero, (6.6) gives the equation for the neutrino. This consideration anticipated the discovery of graphene which came about a decade later. An equation suitable for the neutrino is

$$v_F \vec{\sigma} \cdot \vec{\nabla} \psi(r) = E \psi(r), \tag{6.7}$$

where $v_F \sim 10^6$ m/s is the Fermi velocity which replaces c, the velocity of light; $\psi(r)$ is a wave function of two components; and $\vec{\sigma}$ and E denote Pauli matrices and energy, respectively.

Landau several decades ago argued that such two- and one-dimensional structures would be unstable with their very existence being in question. He was proved to be wrong.

It may be remarked that motion in a two-dimensional sheet is not Lorentz invariant (a theoretical exception maybe a hypothetical infinite sheet). The two-component wave function $\psi(r)$ in (6.7) arises from the wave functions belonging to two side-by-side honeycomb lattices present in graphene. This, in a sense mimics, the spin-up and spin-down of electrons. We will return to this later.

6.5 Graphene and Elementary Particles

We now argue that we can exploit the properties of a graphene sheet to ascertain the behaviour patterns of elementary particles. It may be observed that (6.7) represents the equation for a massless fermion like a neutrino or some quasi-particle. For bi-layered graphene, the mass needs to be considered.

Graphene with its hexagonal structure maybe imagined to be a "chess board". In the sense that chess moves have a minimum length [27]. This minimum length mandates a noncommutative geometry.

So it follows that the commutator is given by

$$[x_i, x_j] = \Theta^{ij} l^2, \tag{6.8}$$

where the noncommutativity of x_i and x_j is evident. This leads to a modification of the Maxwell equations to include an extra term, this was worked out in detail by the author in [28, 29]:

$$\partial^\mu F_{\mu\nu} = \frac{4\pi}{c} j_\nu + A_\lambda \varepsilon F_{\mu\nu}, \tag{6.9}$$

where the symbols have their usual meaning. In Eq. (6.9), ε is a dimensionless number,

$$\varepsilon - 1 \quad \text{for the noncommutative case}$$
$$= 0 \quad \text{otherwise.}$$

With $\varepsilon = 0$ the covariant Maxwell equations are obtained. Let us now consider a two-dimensional case

$$\partial^1 F_{14} = \frac{4\pi}{c} j_4 + A_2 \varepsilon F_{14} \tag{6.10}$$

and such equations for the j_1 and j_2. For the electromagnetic tensor, the equations
are

$$\frac{\partial E_x}{\partial x} = -4\pi \frac{\partial \rho}{\partial t} + \varepsilon A_y E_x \qquad (6.11)$$

$$\frac{\partial E_y}{\partial y} = -4\pi \frac{\partial \rho}{\partial t} + \varepsilon A_x E_y \qquad (6.12)$$

$$-\frac{\partial B_z}{\partial x} = 4\pi j_y + \varepsilon \frac{\partial E_y}{\partial t} \qquad (6.13)$$

$$\frac{\partial B_z}{\partial y} = 4\pi j_x + \varepsilon \frac{\partial E_x}{\partial t}. \qquad (6.14)$$

It must be pointed out that some of these equations are non-steady state. Such fields
could give rise to radiation. Further, if, for example, the electric field gets a sudden
surge due to external factors, then to balance the equation, so should the field on the
left side. This is a separate effect which maybe experimentally observed.

These are in the nature of additional electromagnetic effects. From (6.8), it must
be noticed that there appears to be a magnetic field. This was demonstrated by the
Saito and also the author [30, 31]. Then we have the equation:

$$Bl^2 = \hbar c / e. \qquad (6.15)$$

These considerations are obviously valid for graphene, remembering that constants like v_F and l now have different meanings. In this case,

$$Bl^2 = \hbar v_F / e.$$

The energy for these considerations becomes

$$\text{Energy} = \pm v_F |\vec{p}|.$$

Conduction relates to the positive sign while the negative sign relates to valence
particles. These are the analogues of particles and antiparticles.

The parallel particularly in the Cini–Toushek regime in high-energy physics
is transparent (Cf. Ref. [32]). In the Cini–Toushek regime, a massless feature is
apparent. For such high energies, we have [32]

$$H\psi = \frac{\vec{\alpha} \cdot \vec{p}}{|p|} E(p) \qquad (6.16)$$

bearing a similarity to the massless (6.6). In (6.16), we have

$$\alpha^k = \begin{pmatrix} 0 & \sigma^k \\ \sigma^k & 0 \end{pmatrix} \qquad \beta = \begin{pmatrix} I & 0 \\ 0 & -I \end{pmatrix} \tag{6.17}$$

$$\gamma^0 = \beta \tag{6.18}$$

which can be adapted to the neutrino equation. From (6.8), the Snyder–Sidharth dispersion relation may be obtained as

$$E^2 = p^2 + m^2 + \alpha \frac{l^2}{\hbar^2} p^4 \tag{6.19}$$

α in (6.19) maybe positive or negative.

However, this differs from the Dirac theory—as there is no Lorentz invariance and v_F is an analogue of the velocity of light.

Keeping the above in mind, let us explore the possibility of graphene to be a test bed. The loss of dimensionality has been researched by the author [33, 34] for nearly mono-energetic beams of fermions or bosons. The beam would be two dimensional.

Let us reconsider the occupation number of a fermion gas [17] given by

$$\bar{n}_p = \frac{1}{z^{-1} e^{b E_p} + 1}, \tag{6.20}$$

where $z^{-1} \equiv \frac{\lambda^3}{v} \equiv \mu z \approx z$. Because it can be easily shown $\mu \approx 1$,

$$v = \frac{V}{N}, \lambda = \sqrt{\frac{2\pi \hbar^2}{m/b}}$$

$$b \equiv \left(\frac{1}{KT} \right), \quad \text{and} \quad \sum \bar{n}_p = N. \tag{6.21}$$

Let us consider in particular a collection of fermions which is mono-energetic, in the above context, for which the distribution is

$$n'_p = \delta(p - p_0)\bar{n}_p, \tag{6.22}$$

where $\bar{n}_p$ is given by (6.20) and $\delta(p - p_0)$ is Dirac's Delta function.

Now, again consider a collection of mono-energetic particles in equilibrium.

By the usual formulation, we have

$$
\begin{aligned}
N &= \frac{V}{\hbar^3} \int d\vec{p}\, n'_p \\
&= \frac{V}{\hbar^3} \int \delta(p - p_0) 4\pi p^2 \bar{n}_p dp \\
&= \frac{4\pi V}{\hbar^3} p_0^2 \frac{1}{z^{-1} e^\theta + 1},
\end{aligned}
\tag{6.23}
$$

where $\theta \equiv bE_{p_0}$.

It must be reiterated that in (6.23) there is a loss of dimension in momentum space, because of the presence of the δ function in (6.22).

In an earlier chapter (see Chap. 3), the author showed that neutrinos displayed two-dimensional behaviour. This is expected from the holographic principle [35].

Also, as stated earlier, the universe resembles a black hole, with a black hole being a two-dimensional object [37, 38]. This is because the interior of a black hole is inaccessible. The two dimensionality is because only the area of the black hole may be taken into consideration. The author had shown that the area of a black hole maybe deduced from

$$
A = N l_p^2.
\tag{6.24}
$$

These involve quantum gravity considerations which take into account the quantum of area [38, 39]. That is, the black hole consists of N quanta of area. These quantum gravity features are suitable to be tested by two-dimensional surfaces such as graphene.

In a communication, the author [1] showed for the one-dimensional case, corresponding to nanotubes that

$$
kT = \frac{3}{5} k T_F,
\tag{6.25}
$$

where T_F is the Fermi temperature. For the two-dimensional case also kT is very small. This results from the well-known formulae for two dimensions [20]:

$$
kT = \frac{e\hbar\pi}{m v_F}
\tag{6.26}
$$

$$
(kT)^3 = \frac{6 e \hbar v_F}{\pi},
\tag{6.27}
$$

whence we have

$$
(kT)^2 = 6 \cdot v_F^2 \pi^2 m.
\tag{6.28}
$$

As $v_F \sim 10^8$, kT is very small even for a particle which has an electron-type mass from (6.28). For the Fermi temperature,

$$kT_F = \frac{\hbar}{2}(z \times 6\pi)^{1/3} \cdot v_F$$

just to compare. Another interesting result may be stated, we have

$$v_F^2 = \left(\frac{\hbar\pi}{m}\right)^2 \cdot \frac{1}{A}, \tag{6.29}$$

where $A \sim l^2$ is the quantum of area. So

$$\frac{m^2 v_F^2}{\hbar^2} \cdot l^2 \sim \mathcal{O}(1). \tag{6.30}$$

This agrees with $v_F \to$ the velocity of light c and $\hbar/m v_F \to$ the Compton wavelength. What we would thus like to reiterate is that all that is needed is an ∞ graphene sheet and this leads to the spacetime of relativity and quantum mechanics. In reality, this needs to be a very large sheet of graphene. That is, whatever the temperature, the ensemble effectively mimics a very low-temperature gas. This gives rise to multiple ramifications especially in the context of magnetism.

Thus, we can analyse magnetism and electromagnetism in this new noncommutative perspective. This yields fresh perspectives like the Haas–Van Alphen-type effect [28]. In this latter situation, the magnetization per unit volume shows a periodic behaviour.

6.6 Noncommutative Repercussions

Fluctuations of the *zero point energy* have been widely studied. Based on this the author in 1997, as noted earlier, predicted a contra model of the universe [4, 38] in which there would be a small cosmological constant, that is, an accelerating universe. In 1998, observations of Perlmutter, Reiss, and Schmidt confirmed this scenario. Today we call this dark energy. A manifestation of this is a noncommutative spacetime given in (6.8). This led to the so-called Snyder–Sidharth dispersion relation given in (6.19). We would like to point out that the extra magnetic effect in equations like (6.11) (and the following) can be attributed to this zero point effect of noncommutativity as given in (6.15). Closely related is the Casimir effect which has been observed even in graphene [40, 41]. This is a *zero point energy* fluctuation effect. The Casimir energy in graphene is given by

$$\frac{Energy}{area} = \frac{\pi^2}{240} \cdot \frac{\hbar c}{a^3}. \tag{6.31}$$

The energy itself is given by

$$Energy = \left(\frac{\pi^2}{240}\right) \cdot \frac{\hbar c}{a},$$ (6.32)

where we consider the area to be $\sim a^2$.

If, following Wheeler [14], we consider directly ground state oscillators of the *zero point energy*, we can deduce that

$$Energy \sim \hbar c / a$$

resembling (6.32). Similarly if we take the extra term in the dispersion relation (6.19), it is easy to show that this also has the same form. All this is hardly surprising because they are all manifestations of fluctuations in the quantum vacuum.

It must be mentioned that the Casimir effect in graphene has been observed (see [42]). What is interesting is that a group of scientists from MIT, Harvard University, Oak Ridge National Laboratory, and other universities have used this *zero point energy* for a compact integrated silicon chip. Clearly the same would be possible for graphene too particularly in the context of quantum computers: The "spin"-up and spin-down being the qubits [43].

To proceed further we invoke (6.15) and the well-known result for a coil

$$\iota = \frac{NBA}{R\Delta t},$$ (6.33)

where N is the number of turns, A is the area, and R is the resistance. Use of (6.15) in (6.33) now gives

$$\iota \approx \frac{NA}{R} \cdot \frac{e}{l^2\tau}.$$ (6.34)

Whatever be N, if we think of a coil made up of nanotubes or graphene, remembering that l is small and so is the resistance (6.34) would be observable, like indeed (6.15).

Further observing that nanotubes and graphene can harbour fast-moving fermions (including neutrons) and of course carbon, we have all the ingredients for manipulating a version of table-top fusion possibly using the bosonization of fermions property. In this case, we use an equation like (6.23) and preceding consideration [1, 38].

To proceed, in this case, $kT = < E_p > \approx E_p$ so that $\theta \approx 1$. But we can continue without giving θ any specific value.

Using the expressions for v and z given in (6.21) in (6.22), we get

$$(z^{-1}e^\theta + 1) = (4\pi)^{5/2}\frac{z^{'-1}}{p_0}; \text{ whence}$$

$$z^{'-1}A \equiv z^{'-1}\left(\frac{(4\pi)^{5/2}}{p_0} - e^\theta\right) = 1,$$ (6.35)

where we use the fact that in (6.21), $\mu \approx 1$ as can be easily deduced.

A number of conclusions can be drawn from (6.35). For example, if,

$$A \approx 1, \qquad i.e.$$

$$p_0 \approx \frac{(4\pi)^{5/2}}{1+e}, \tag{6.36}$$

where A is given in (6.35), then $z' \approx 1$. Remembering that in (6.21), λ is of the order of the de Broglie wavelength and v is the average volume occupied per particle, this means that the gas gets very densely packed for momenta given by (6.36). In fact for a Bose gas, as is well known, this is the condition for Bose–Einstein condensation at the level $p = 0$ [17].

In any case, there is an anomalous behaviour of the fermions.

6.7 Two-Dimensional Considerations

Some enigmatic features of graphene may be accounted for by invoking noncommutative geometry generated by the lattices of graphene which are like a honeycomb (or any two-dimensional crystal), for example, properties like the minimum conductivity or the mysterious fractional quantum Hall effect. The authors in reality in [27] study a model of electrons leap-frogging from atom to atom in graphene's honeycomb lattice. They claim that this jumping gives rise to low-energy electronic excitations which can be identified with a 2+1-dimensional Dirac equation. Graphene's "pseudospin" arises from the degeneracy present in the honeycomb lattice. It is also claimed by them that pseudospin is a real angular momentum. This, they claim, explains features like the suppression of backscattering in carbon nanotubes.

The author has argued that in the infinite limit of these sheets, Minkowski spacetime is recovered, with the lattice length replaced by Compton length [44]. The test bed characteristic of graphene alluded to in Sect. 6.5 could be achieved by "scaling"—say, the velocity of light c replacing the Fermi velocity.

Let us refer to the work of Nobel Laureate, Geim [45], where he studies electron waves propagating through the honeycomb lattices of graphene. Then, these electrons, according to Geim, lose their effective mass, rendering them to be quasiparticles. Such quasi particles maybe represented by a Dirac-like equation rather than the Schrödinger equation. The Schrödinger equation—which is successful in depicting quantum properties of other materials—fails when it comes to graphene's moving charges with their zero rest mass. Just as in [45] we use a two-component Dirac equation as for the massless neutrino

$$\sigma^\mu \partial_\mu \psi = 0 \tag{6.37}$$

the σs being the two component Pauli matrices.

Thence, the author showed some interesting features:

For instance, a noncommutative geometry holds for graphene (Cf. also [28]). The lattices display a homogenous structure which leads to the noncommutative feature. The author independently had shown that noncommutative geometry yields magnetic fields. Saito [31] through a different approach also came to the same conclusion [46]. The author derived a relation for the magnetic field:

$$Bl^2 = \hbar c/e, \tag{6.38}$$

which has been seen in the author's work before. This can be worked with more detail using the connectedness of non-integrable space [30]. The starting point is a non-integrable infinitesimal parallel displacement of a four vector:

$$\delta a^\sigma = -\Gamma^\sigma_{\mu\nu} a^\mu dx^\nu. \tag{6.39}$$

The Γ's are the Christoffel symbols, representing the extra effect in displacements, because of curved space. If space were flat the Γ's on the right-hand side would disappear. Differentiating partially with respect to the μ-th coordinate, this gives, from (6.39),

$$\frac{\partial a^\sigma}{\partial x^\mu} \rightarrow \frac{\partial a^\sigma}{\partial x^\mu} - \Gamma^\sigma_{\mu\nu} a^\nu. \tag{6.40}$$

The second term on the right side of (6.40) can be written as

$$- \Gamma^\lambda_{\mu\nu} g^\nu_\lambda a^\sigma = -\Gamma^\nu_{\mu\nu} a^\sigma.$$

When the metric is linearized we have

$$g_{\mu\nu} = \eta_{\mu\nu} + h_{\mu\nu},$$

$\eta_{\mu\nu}$ being the Minkowski metric and $h_{\mu\nu}$ a small correction whose square is neglected. From (6.40), we conclude that

$$\frac{\partial}{\partial x^\mu} \rightarrow \frac{\partial}{\partial x^\mu} - \Gamma^\nu_{\mu\nu}. \tag{6.41}$$

We can identify

$$A_\mu = \Gamma^\nu_{\mu\nu} \tag{6.42}$$

exactly as in Dirac's monopole theory from the above, using minimum electromagnetic coupling.

If we use (6.41), we obtain the commutator relation:

$$\frac{\partial}{\partial x^\lambda} \frac{\partial}{\partial x^\mu} - \frac{\partial}{\partial x^\mu} \frac{\partial}{\partial x^\lambda} \rightarrow \frac{\partial}{\partial x^\lambda} \Gamma^\nu_{\mu\nu} - \frac{\partial}{\partial x^\mu} \Gamma^\nu_{\lambda\nu}. \tag{6.43}$$

Let us now use (6.42) in (6.43): The R. H. S. does not vanish due to the presence of the electromagnetic field (6.42) and the momentum components of quantum theory being noncommutative. Indeed the L.H.S. of (6.43) can be written as

$$[p_\lambda, p_\mu] \approx \frac{\mathcal{O}(1)}{l^2},$$ (6.44)

l being the Compton wavelength or minimum length. In (6.44), we have brought into play the fact that at the extreme scale of the Compton wavelength, the momentum is mc (the Planck scale being a special case). For graphene this translates to $m \times$ the Fermi velocity.

From (6.42), (6.43), and (6.44), we have

$$Bl^2 \sim \frac{1}{e} \left(= \frac{\hbar c}{e} \right),$$ (6.45)

where B is the magnetic field, with $\hbar$ and c restored from the natural units. There is an alternative method of proving (6.45). From the Landau theory of synchroton radiation, the frequency ω is given by

$$\omega = eB/mc.$$ (6.46)

And the maximum value of ω is given by [38]

$$\omega = c/l.$$ (6.47)

Using (6.47) in (6.46), we get (6.45) (Cf. also [47]).

6.8 Electromagnetic Effects Due to Graphene

In order to understand the electromagnetic effects due to graphene, let us take the inter-lattice distance to be l. It must be borne in mind that the magnetic field B and the electric field E are generated by the geometry of the system. The minimum conductivity of graphene has been regarded as a puzzling feature which maybe attributed solely to the geometry of the system, rather than due to any external influences (Cf. Ref. [44]).

Further it was shown by the author [44] that if an infinite graphene sheet is chosen, it displays properties of Minkowski space including noncommutativity. Here the lattice constant $L \sim 2\mathring{A}$ replaces the Compton wavelength. To see this, we note that

$$c = 300v_F \quad \text{and} \quad m_v \sim 0.05\, m,$$ (6.18)

where m_ν is the graphene "electron" mass and m is the electron mass. Substituting this into the Compton wavelength expression

$$l = \frac{\hbar}{mc}, \tag{6.49}$$

where l bears a strong resemblance to L or correspond to each other. Here, we can immediately recognize from (6.48) and (6.49) a Reynold number-type scaling relation.

Thus, it appears that B and E which arise in graphene due to the geometry of the structures could equally well be applicable to the Minkowski space in general after due scaling.

An interesting observation here is that in the case of bi-layer graphene, there would be a small mass, which would lead to an identification with the four-component Dirac equation [48], rather than the earlier bispinorial case

$$(\partial_\mu \gamma^\mu - m)\psi = 0, \tag{6.50}$$

where γs are the Dirac four-component matrices.

So there would be no chirality now, but noncommutative geometry would still be applicable. In this case, the above considerations would be approximately valid.

On the other hand, in an earlier era, the origin of electromagnetism in statistical physics was explained via the Bohr–van Leeuwen theorem which went as follows. *When statistical mechanics and classical mechanics are applied consistently, the thermal average of the magnetization is always zero.*

6.9 The Fractional Quantum Hall Effect in Relation with Graphene

We propose to use (6.45) to derive the otherwise inexplicable fractional quantum Hall effect [49]:

$$BL^2 = \hbar c/e, \tag{6.51}$$

where L^2 defines a quantum of area exactly as in quantum gravity approaches [28, 29]. This is the area of individual lattices, in our case.

In these considerations as noted, the Fermi velocity v_F replaces the velocity of light. So we have for the electron mobility and conductivity

$$\mu = v_F/|E| \tag{6.52}$$

$$\sigma = (n/A)e \cdot \frac{v_F}{|E|}, \quad A \sim L^2, \tag{6.53}$$

where A, as in the usual theory, is the area and n is the number of electrons. In our case as noted above A, the area is made up of a number of honeycomb lattice areas, each with area $\sim L^2$, that is,

$$A = mL^2,$$

where m is an integer.

We also note that the electric field strength E equals the magnetic field strength B in the case of 2D structures (Cf. Ref. [49]). Using these inputs in (6.53) we get

$$\sigma = \frac{n}{m} \cdot \frac{ev_F}{|B|L^2}. \tag{6.54}$$

If we now use (6.51) in (6.54) (with v_F replacing c) we get for the conductance

$$\sigma = \frac{n}{m} \cdot \frac{e^2}{h} \tag{6.55}$$

which defines the fractional quantum Hall effect.

Earlier the author had shown that it is this noncommutative space feature in two-dimensional structures that explains also Landau levels [47] or the minimum conductivity that exists in graphene even when there are practically no electrons at the Dirac points [50] (Cf. also Ref. [44]). In other words, several supposedly diverse phenomena arise from the noncommutative space of these two-dimensional structures and by extension, in Minkowski spacetime, e.g. the origin of electromagnetism itself.

6.10 The Conductivity Mystery with Graphene

One of the mysteries involving graphene is that there is a minimum conductivity which does not disappear. This conductivity (as seen above) is given by

$$\sigma = 4e^2/\hbar. \tag{6.56}$$

However, this mystery is easily solved if we remember that as seen earlier, because there is noncommutativity of space in the hexagonal two-dimensional crystal. We have as seen from Eqs. (6.11)–(6.14) that there is an extra electric and magnetic fields. To proceed, we have for the electron mobility μ and conductivity σ:

$$\mu = v_F/|E| \tag{6.57}$$

$$\sigma = (n/A)e \cdot \frac{v_F}{|E|}, \quad A \sim l^2. \tag{6.58}$$

The minimum of $n \sim 1$. Whence we get

$$\sigma = \frac{1}{l^2} \cdot \frac{e \cdot v}{|B|} \text{ where } l = \left(\frac{\hbar}{2\,m v_F} \right). \tag{6.59}$$

From (6.59) we can easily obtain (6.56). In other words, the mysterious minimum conductivity is due to the extra magnetic effect of noncommutative spacetime which holds for graphene.

If we consider in (6.58) n to be the number of electrons in general and A to be the area of m honeycomb lattices, then we can get from (6.58) the fractional quantum Hall effect. It would be interesting to consider a computer with a two-dimensional chip that would replicate some of the above features.

6.11 Magnetic Effect of Noncommutativity Again

We know that the relativistic energy–momentum relation is given by

$$p_\mu p^\mu = \frac{E^2}{c^2} - p^2 = m^2 c^2 \tag{6.60}$$

or, equivalently,

$$E^2 = p^2 c^2 + m^2 c^4. \tag{6.61}$$

Now if we take natural units, i.e. $\hbar = c = 1$ we can rewrite this equation as

$$E^2 = p^2 + m^2. \tag{6.62}$$

This Eq. (6.62) can be considered only in the case where spacetime is continuous. But if we consider noncommutative geometry and the Snyder relation of position and momentum [51–53]

$$[x, p] = \hbar = \hbar[1 + (l/\hbar)^2 p^2], \tag{6.63}$$

where l is the minimum length. We can see that if $l \to 0$ we get back the usual Heisenberg relation of position and momentum. Substituting Eq. (6.63) in (6.62), we get

$$E = (m^2 + p^2[1 + l^2 p^2]^{-2})^{1/2} \tag{6.64}$$

or

$$E^2 = p^2 + m^2 + \alpha l^2 p^4. \tag{6.65}$$

As can be seen this equation is the so-called Snyder–Sidharth dispersion relation for fermions. This relation becomes important at high energies, for example, at energies which are expected in LHC [38].

6.12   The Honey-Comb Structure of Two-Dimensional Graphene

We can now say that it is the noncommutative spacetime that leads to magnetic effects. A recent description of spacetime in graphene was given in [27] which approximates a 2D scenario for graphene. The authors show that the spin of the electron itself is due to the discrete nature of spacetime. Spacetime is considered by these authors to resemble a chess board. Such a discrete partition of spacetime would lead to the generation of a magnetic field. This magnetic field could affect the behaviour of the electron.

The author has argued that spacetime is discrete and that during the hopping of electrons they could undergo a change in spin direction. The author has also postulated that space is not smooth. This gives rise to a magnetic field which can cause the change of spin in the electron during hopping. So we can see that the changing of the electron spin maybe compared with the appearance of a magnetic field in noncommutative spacetime.

At this juncture, it maybe relevant to talk about the Hall effect. We comment on the Hall effect from the following novel point of view, namely, the strong parallel between two-dimensional quantum mechanics and graphene [44, 54]. The Hall effect itself was observed in the nineteenth century, in the case of a current (or a moving electron) I_x along the x-axis, a magnetic field B_z along the z-axis, leading to the Hall voltage along the y-axis. This is given by

$$V_H = \left(\frac{1}{n|e|}\right) \cdot \frac{I_x}{d} B_z. \tag{6.66}$$

The expression $\left(\frac{1}{n|e|}\right)$ is called the Hall resistance, V_H being the Hall voltage, and d the thickness of the conductor. (The Hall effect, as known, has a parallel in relativistic electromagnetic theory.)

Electrons exhibit strange neutrino-like properties, in two and one dimensions as pointed out by the author, from the mid-90s [2, 3, 25] (graphene and nanotubes). The two component equation [26] for this is

$$\left(\sigma^\mu \partial_\mu - \frac{mc}{\hbar}\right) \psi = 0, \tag{6.67}$$

where σ^μ denote the 2×2 Pauli matrices. In case the mass vanishes, (6.67) gives the neutrino equation. This is pertinent in the case of graphene. Incidentally, graphene was discovered nearly a decade later. For the electron quasi-particles in graphene, we have

$$v_F \vec{\sigma} \cdot \vec{\nabla} \psi(r) = E \psi(r), \tag{6.68}$$

where $v_F \sim 10^6$ m/s is the Fermi velocity which comes in place of c, the velocity of light and $\psi(r)$ stands for a two-component wave function, E denotes energy.

Indeed, taking this to the next level, the author has pointed out that graphene (or more generally two-dimensional structures) could be a candidate for a test bed for high-energy physics. This may be understood in the sense of the role played by a wind tunnel, knowing Reynold's numbers, for the actual problem [44, 54]. With this input, we can solve problems like the minimum conductivity observed in graphene, as also throw light on the fractional quantum Hall effect, something that has escaped explanation.

It maybe noted that relativistic effects are exhibited by graphene when the Fermi velocity v_F replaces the velocity of light. And this also gives a "Lorentz" transform. Furthermore, in the electromagnetic case, this leads to the Lorentz force equal to $\vec{v} \times \vec{B}$ where $\vec{v}$ is the velocity of the moving or conduction electron and $\vec{B}$ is the magnetic field. This Lorentz force can be immediately identified with the Hall effect EMF.

The Lorentz force may be written as

$$\text{Force} = \frac{d\vec{p}}{dt} = \frac{e}{c}\vec{u} \times \vec{B}. \tag{6.69}$$

So the energy is given by

$$\text{Energy} = \frac{1}{c} R \frac{I_x B_z}{d}, \tag{6.70}$$

where a factor R denotes resistance, for the special case when the electron is not a free moving particle. Comparing (6.66) and (6.70) it is apparent from here that the Hall effect corresponds to the Lorentz force in relativistic electrodynamics.

It must also be pointed out that as shown in [44, 54], it is the noncommutative nature of the graphene space which leads to enigmatic properties like the minimum conductance in graphene or the fractional quantum Hall effect.

There are two other pertinent facts which may be emphasized. The first is that the magnetic field here is stronger than the usual Maxwellian field. In noncommutative space, it maybe written as

$$Bl^2 = \frac{\hbar c}{e}. \tag{6.71}$$

In Eq. (6.71), symbols have their usual meaning except for l which is the minimum length, being the lattice length here. This was deduced independently by the author and partly by Saito (independently) some years ago [46]. The experimentally observed and puzzling minimum conductivity is given by

$$\sigma = 4\frac{e^2}{\hbar}, \tag{6.72}$$

with the symbols having their usual meanings in Eq. (6.72). What is remarkable is that the magnetic field (6.71) and the electric current following from (6.72) arise only because of the noncommutativity of the spacetime in these two-dimensional structures.

6.13 Graphene and Entanglement

We now consider entanglement in 2D structures. For a 2D crystal-like graphene, let us consider an Ising-like lattice model. In case of lattice models, the Hamiltonian from mean field theoretic approach is known to be given by

$$H = H_0 + H_1, \tag{6.73}$$

where $H_0 = -\sum_{(s_1,s_2)} E\phi_{s1}\phi_{s2}$, $H_1 = -\sum_s F\phi_s$, E and F being the energy and vacuum field energy respectively, and ϕ_s is the state variable. It can be shown that in the absence of fluctuations, the corresponding energy for a single lattice site ($s \in S$) must be greater or equal to the average value of all corresponding energy configurations for that site. Essentially this means

$$F\phi_s \geq\ <F\phi_s>. \tag{6.74}$$

Thus, considering that there are no fluctuations and the condition (6.74) is fulfilled, one can visualize a graphene sheet of length l and area a that has the Hamiltonian as (6.73). Now, since graphene is known to have a minimum conductivity. We could investigate if the phenomena of entanglement can be manipulated using the minimal conductivity and the mean field theoretic approach. For more details on this, see [55].

References

1. B.G. Sidharth, Anomalous Fermions. J. Stat. Phys. **95**(3), 775–784 (1999)
2. B.G. Sidharth, A note on Two Dimensional Fermions. *BSC-CAMCS-TR-95-04-01* (1995)
3. B.G. Sidharth, A Note on Degenerate and Anamolous Bosons (1995). https://arxiv.org/abs/quant-ph/9506002
4. B.G. Sidharth, The universe of fluctuations. Int. J. Mod. Phys. A **13**(15), 2599–2612 (1998)
5. B.G. Sidharth, Quantum mechanical black holes: towards a unification of quantum mechanics and general relativity. Indian J. Pure Appl. Phys. **35**, 456 ff (1997)
6. B.G. Sidharth, Quantum Black Holes, in *Frontiers of Quantum Physics-1997*. ed. by C.H. Lim (Springer, Singapore, 1998), pp. 308–309
7. B.G. Sidharth, Quantum nonlinearities in the context of black holes. Nonlinear World **4**, 157–162 (1997)

8. P.A.M. Dirac, *The Principles of Quantum Mechanics*, vol. 27 (Oxford university press, 1981)
9. B.G. Sidharth, Large numbers and the time variation of physical constants. Int. J. Theor. Phys. **37**(4), 1307–1312 (1998)
10. S. Perlmutter, Discovery of a supernova explosion at half the age of the Universe. Nature **391**(6662), 51–54 (1998)
11. J. Glanz, Astronomy: cosmic motion revealed. Science **282**(5397), 2156–2157 (1998). https://www.science.org/doi/abs/10.1126/science.282.5397.2156a
12. B.G. Sidharth, Quarks or Leptons? Mod. Phys. Lett. A **12**(32), 2469–2471 (1997)
13. P. Schiffer, D.D. Osheroff, Nucleation of the AB transition in superfluid He_3: surface effects and baked Alaska. Rev. Mod. Phys. **67**(2), 491 (1995)
14. C.W. Misner, K.S. Thorne, *and J* (Freeman, A. Wheeler. Gravitation, 2000)
15. J.D. Bjorken, S.D. Drell, *Relativistic Quantum Mechanics* (Mcgraw-Hill College, 1964)
16. A. Zee, *Unity of Forces in the Universe*, vol. 2 (World Scientific, 1982)
17. K. Huang, *Statistical Mechanics* (Wiley Eastern, New Delhi, 1975)
18. R. Sridhar, *Advances in Theoretical Physics* (Pathak AP, Narosa, New Delhi, 1996)
19. D.C. Lieb, E.H. Mattis. *Mathematical Physics in One Dimension: Exactly Soluble Models of Interacting Particles* (Academic, 2013)
20. F. Reif, *Fundamentals of Statistical and Thermal Physics* (Waveland Press, 1965)
21. P. Delaney, H.J. Choi, J. Ihm, S.G. L, M.L. Cohen, Broken symmetry and pseudogaps in ropes of carbon nanotubes. Nature **391**(6666), 466–468 (1998)
22. M.S. Dresselhaus, New tricks with nanotubes. Nature **391**(6662), 19–20 (1998)
23. J. Wilder, L.C. Venema, A.G. Rinzler et al., Electronic structure of atomically resolved carbon nanotubes. Nature **391**(6662), 59–62 (1998)
24. T.W. Odom, J. Huang, P. Kim, C.M. Lieber, Atomic structure and electronic properties of single-walled carbon nanotubes. Nature **391**(6662), 62–64 (1998)
25. B.G. Sidharth, Low dimensional electrons, in *Solid State Physics*, ed. by R. Mukhopadhyay et. al, vol. 41 (1999), p. 331
26. B.G. Sidharth, *Chaotic Universe* (Nova Science, New York, 2001)
27. M. Mecklenburg, B.C. Regan, Spin and the honeycomb lattice: lessons from graphene. Phys. Rev. Lett. **106**, 116803 (2011). (Mar)
28. B.G. Sidharth, Magnetic effect of noncommutativity. Italian J. Pure Appl. Math. 493 (2013)
29. B.G. Sidharth, The common origin of the cosmological constant and the higgs mechanism. Int. J. Theor. Phys. **52**, 1532–1538 (2013)
30. B.G. Sidharth, The elusive monopole. Nuovo Cimento B Serie **118**(1), 35 (2003)
31. T. Saito, Gravitation Cosmol. **6**(22), 130–136 (2000)
32. B.G. Sidharth, High energy Dirac solutions: Issues and ramifications. Int. J. Theor. Phys. **53**(5), 1561–1568 (2014)
33. B.G. Sidharth, Fractal statistics. Chaos, Solitons & Fractals **12**(13), 2475–2480 (2001)
34. B.G. Sidharth, Consequences of a quantized spacetime model. Chaos, Solitons & Fractals **13**(4), 617–620 (2002)
35. B.G. Sidharth, A model for neutrinos. Int. J. Theor. Phys. **52**, 4412–4415 (2013)
36. A.D. Popova, B.G. Sidharth, Is the Universe asymptotically spatially two dimensional. Nonlinear World **4**, 493–504 (1997)
37. B.G. Sidharth, A note on the cosmic neutrino background and the cosmological constant. Found. Phys. Lett. **19**(7), 757–759 (2006)
38. B.G. Sidharth, *The Thermodynamic Universe* (World Scientific, Singapore, 2008)
39. J. Baez, The quantum of area? Nature **421**(6924), 702–703 (2003)
40. I.V. Fialkovsky, V.N. Marachevsky, D.V. Vassilevich, Finite-temperature Casimir effect for Graphene. Phys. Rev. B **84**(3), 035446 (2011)
41. M. Bordag, I.V. Fialkovsky, D.M. Gitman, D.V. Vassilevich, Casimir interaction between a perfect conductor and Graphene described by the Dirac model. Phys. Rev. B **80**(24), 245406 (2009)
42. G.L. Klimchitskaya, U. Mohideen, V.M. Mostepanenko, The Casimir effect in graphene systems: experiment and theory. Int. J. Mod. Phys. A **37**(19) (2022)

43. J. Zou, Z. Marcet, A.W. Rodriguez et al., Casimir forces on a silicon micromechanical chip. Nat. Commun. **4**(1), 1–6 (2013)
44. B.G. Sidharth, Graphene and high energy physics. Int. J. Mod. Phys. E **23**(05), 1450025 (2014)
45. A.K. Geim, Graphene: status and prospects. Science **324**(5934), 1530–1534 (2009)
46. B.G. Sidharth, Fuzzy noncommutative space–time: a new paradigm for a new century, frontiers of fundamental physics (2001)
47. B.G. Sidharth, A test for noncommutative spacetime. Int. J. Mod. Phys. E **14**(08), 1247–1249 (2005)
48. W. Greiner, B. Muller, J. Rafelski, *Quantum Electrodynamics of Strong Fields* (Springer, 1985)
49. B.G. Sidharth, A brief note on the fractional Quantum Hall effect. Int. J. Theor. Phys. **54**(7), 2382–2385 (2015)
50. B.G. Sidharth, Dark energy and graphene. New Adv. Phys. **8**(1), 29–50 (2014)
51. H.S. Snyder, The electromagnetic field in quantized spacetime. Phys. Rev. **72**(1), 68 (1947)
52. H.S. Snyder, Quantized spacetime. Phys. Rev. **71**(1), 38 (1947)
53. B.G. Sidharth, High Energy P-P Collisions. Int. J. Mod. Phys. E **19**(10), 2023–2031 (2010)
54. B.G. Sidharth, J. Rishab, Graphene possibilities. New Adv. Phys. **8**(2), 159–162 (2014)
55. B.G. Sidharth et al., Entanglement and it's manipulation. Int. J. Theor. Phys. **58**, 2936–2941 (2019)

Chapter 7
A Fifth Force in Nature?

7.1 Introduction

The prevalent theory among physicists about the fundamental interactions has been that the four basic interactions were adequate to explain all known phenomena. These interactions, as is known, are gravitation, electromagnetism, electroweak force, and what were previously called strong interactions. However, over two decades ago, the author, in a talk at Vanderbilt University, USA proposed that there is a brand new force over and above the four well-known forces. His model was that of a proton and a neutron in the nucleus which, in the absence of other known forces, would be identical particles. It is only when the electromagnetic interactions are "switched on" that they appear to be distinct. This proposal was reported in the press. It may be pertinent to mention that the author had proposed his theory of dark energy way back in 1997 which was observationally confirmed the very next year [3].

As far as the fifth force is concerned, this was already hinted by the author's conceptualization of elementary particles as described by the Kerr–Newman metric. There have been claims off and on about the detection of a new and fifth force, notably from Fermilab, Chicago, and a Hungarian team [4, 5]. These claims were also substantiated by the University of California, Irvine. And yet a sense of scepticism persists among physicists. Recently, the LHC b or the Large Hadron Collider b in CERN, Geneva has given a further fillip to the existence of a fifth force. The team in LHC b was working on the beauty quark and its decay.

B. G. Sidharth, *The Dark Energy Paradigm*,
https://doi.org/10.1007/978-981-96-3745-4_7

7.2   Theoretical Justification for the Existence of a Fifth Force

1. The Kerr–Newmann Quantum Mechanical Black Hole

When distances in the range of the Compton wavelength are focussed upon, many quantum phenomena come to light. These include phenomena like the *zitterbewegung*, energy solutions which are negative and luminal velocities for particles not limited to photons. We start with the Kerr–Newman metric in which both quantum mechanics and general relativity are interlaced. This leads to a depiction of leptons and quarks. The Kerr–Newmann metric is known to be represented as follows (cf. Ref. [7]):

$$ds^2 = - \frac{\Delta}{\rho^2}[dt - a\sin^2\theta d\phi]^2$$

$$+ \frac{\sin^2\theta}{\rho^2}[(r^2 + a^2)d\phi - adt]^2$$

$$+ \frac{\rho^2}{\Delta}dr^2 + \rho^2 d\theta^2, \tag{7.1}$$

where a is the Compton wavelength and

$$\Delta = r^2 - 2mr + a^2 + m^2 + e^2, \; \rho^2 \equiv r^2 + a^2\cos^2\theta.$$

At $r = a$ and $\theta = \pi/2$, $\Delta = 2a^2$ as both e and $m \ll a$, and $\rho^2 = a^2$.
We note here from Eq. (7.1) that there is additionally a short-range force of the order of $1/r^3$. There are also other forces of the order of $1/r^3$. These are shorter and stronger as r becomes small.

2. Vector bosons with a short range and life

Let us write the Dirac equation [8, 9] as

$$(\gamma^\mu p_\mu - m)\psi = 0. \tag{7.2}$$

The γ^μ are the usual 4×4 matrices. The algebra they follow is the Clifford algebra. Here ψ is a four-component spinor given by

$$\psi = \begin{pmatrix} \phi \\ \chi \end{pmatrix}. \tag{7.3}$$

In Eq. (7.3) ϕ and χ are two-component spinors. ϕ is sometimes called the large component. It is the positive-energy two component of (7.3). Similarly, χ of (7.3) is the small or negative component. Furthermore, it is known that

$$\chi \sim \left(\frac{v}{c}\right)^2 \phi. \tag{7.4}$$

However, at very high velocities (near the luminal velocity) ϕ and χ reverse their roles (Cf. Refs. [10, 11]) and maybe described by the following equations:

$$\imath\hbar(\partial\phi/\partial t) = c\tau \cdot (p - e/cA)\chi + (mc^2 + e\phi)\phi,$$

$$\imath\hbar(\partial\chi/\partial t) = c\tau \cdot (p - e/cA)\phi + (-mc^2 + e\phi)\chi. \tag{7.5}$$

If there is no electromagnetism then from (7.5) we get

$$t \to -t, \quad \phi \to -\chi. \tag{7.6}$$

This in particular is true at or very near near the Compton scale. Furthermore, at this scale, we encounter *zitterbewegung* and other such phenomena. We can argue, as can be easily seen, also that time can be described by a [1] double Wiener process, then we can write

$$\frac{d_+}{dt}x(t) = \mathbf{b}_+, \quad \frac{d_-}{dt}x(t) = \mathbf{b}_-. \tag{7.7}$$

In the one-dimensional case, for example. Equation (7.7) shows that $x(t)$ is not differentiable at time t. We must remember that without loss of generality and for simplicity we are considering the one-dimensional case. Now let us consider the Fokker–Planck equations [12, 13]

$$\partial\rho/\partial t + div(\rho\mathbf{b}_+) = V\Delta\rho,$$

$$\partial\rho/\partial t + div(\rho\mathbf{b}_-) = -U\Delta\rho, \tag{7.8}$$

where
$$V = \frac{\mathbf{b}_+ + \mathbf{b}_-}{2}; \quad U = \frac{\mathbf{b}_+ - \mathbf{b}_-}{2}. \tag{7.9}$$

Whence we can easily deduce

$$\partial\rho/\partial t + div(\rho V) = 0 \tag{7.10}$$

$$U = v\nabla ln\rho. \tag{7.11}$$

In (7.10) and (7.11), V and U are averages of the respective velocities and the velocity differences.
We define the following
$$V = 2v\nabla S \tag{7.12}$$

$$V - \imath U = -2\imath v\nabla(ln\psi). \tag{7.13}$$

The significance of all this will be seen later (see [14] for further details). We now point out what maybe loosely called the complimentary roles of position and velocity: the complex velocity in (7.13) is defined for point space in terms of a positive time or time flowing in the usual direction where the coordinates are complex.

$$x \to x + \iota x'. \tag{7.14}$$

Further (7.9) maybe written as

$$\frac{dX_r}{dt} = V, \quad \frac{dX_\iota}{dt} = U. \tag{7.15}$$

Here X is complex coordinate with real and imaginary parts X_r and X_ι. On the other hand, the coordinate X maybe differentiated with respect to time, where we treat time as the usual uni-directional coordinate.
We can deduce from (7.9) and (7.15) that

$$W \equiv \frac{d}{dt}(X_r - \iota X_\iota) \tag{7.16}$$

(Cf. Ref. [15]).
Let us now cross over (7.14), which is in one dimension to three dimensions. This leads us surprisingly to four dimensions:

$$(1, \iota) \to (I, \tau),$$

where I is the unit 2×2 matrix and τs are the Pauli matrices. Remarkably this yields the Lorentz invariant metric. What this shows is

$$x + \iota y \to I x_1 + \iota x_2 + j x_3 + k x_4. \tag{7.17}$$

In this equation (ι, j, k) are the Pauli matrices. This leads to

$$x_1^2 + x_2^2 + x_3^2 - x_4^2 \tag{7.18}$$

which is invariant. What we have done is, demonstrated the existence of a one-to-one correspondence between (7.17) and Minkowski four vectors as shown by (7.18).
Whence, it is easy to show that

$$[x^\iota \tau^\iota, x^j \tau^j] \propto \varepsilon_{\iota j k} \tau^k \neq 0 \tag{7.19}$$

follows. Here ι or j are dummy indices as far as the summation is concerned. The whole exercise is to show that special relativity can follow from the above considerations. Alternatively, absorbing the x' and τ' into each other, following

Feshbach and Villars [10], (7.19) transforms into

$$[x^i, x^j] = \beta \varepsilon_k^{ij} \tau^k. \tag{7.20}$$

In Eqs. (7.19) and (7.20), the commutator brackets are $\neq 0$ and this obviously implies that the coordinates do not commutate. The author has been working on this type of spacetime (7.19) with the underlying noncommutativity in (Cf. [12]). It maybe recalled that Snyder had introduced such considerations based on the existence of a minimum spacetime length. It was hoped at that time that this would circumvent the problems which arose in quantum field theory like divergences. As we are considering only a positive-energy picture of the cosmos, we take into account the Compton wavelength to be the fundamental length. This is because, otherwise within the Compton wavelength we encounter several novel but unphysical phenomena like negative energies, *zitterbewegung*, and so on [16–18].

We return at this stage to the Feshbach and Villars description in the context of these considerations. Reverting to (7.3), to return to the description of a particle–antiparticle pair.

Utilizing the $SU(2)$ group representation we finally get [19]

$$\psi(x) \to exp[\frac{1}{2}ig\tau \cdot \omega(x)]\psi(x). \tag{7.21}$$

We can observe that Eq. (7.21) yields a gauge covariant derivative as follows:

$$D_\lambda \equiv \partial_\lambda - \frac{1}{2}ig\tau \cdot \overline{W}_\lambda. \tag{7.22}$$

From here, vector bosons $\overline{W}_\lambda$ can be obtained. Further, this in turn leads to a weak-type interaction:

$$\overline{W}_\lambda \to \overline{W}_\lambda + \partial_\lambda \omega - g\omega \Lambda \overline{W}_\lambda. \tag{7.23}$$

Nevertheless, we must bear in mind that we are no longer dealing with the usual isospin, but rather the interaction among positive and negative energy states (cf. Eq. (7.5)), that is, with particles and antiparticles. Here, we would like to point out that this interaction throws up a new *non-electroweak force* between particles and antiparticles. This force would have a short life, being as it is at the Compton scale. The Klein–Gordon equation description due to Feshbach and Villars would also have these features [10, 11]. Equations like (7.5) would be derivable, the only difference being that ϕ and χ are scalar functions. It must be reiterated for emphasis that the usual picture involving positive-energy solutions is valid only above the Compton scale (Cf. Refs. [8, 9]). Thus, in other words, Eq. (7.3) throws up a new spinor in what maybe labelled "superspin" space.

All this shows that vector bosons $\overline{W}$ encountered here, describe a new *short-range force.*

3. Let us now recall the following

$$g_{\mu\nu} = \eta_{\mu\nu} + h_{\mu\nu}, \quad h_{\mu\nu} = \int \frac{4T_{\mu\nu}(t - |\vec{x} - \vec{x}'|, \vec{x}')}{|\vec{x} - \vec{x}'|} d^3x', \tag{7.24}$$

where

$$T^{\mu\nu} = \rho u^\mu u^\nu. \tag{7.25}$$

The author had shown earlier [12] that (7.25) in (7.24) generates spin, gravitational potential and charge of an electron. This of course is at the Compton scale (Cf. [2] for details). We now invoke the macro-gravitoelectric and gravitomagnetic equations. This leads to the following: (Cf. Ref. [21]).

$$\nabla \cdot \vec{E}_g \approx -4\pi\rho, \quad \nabla \times \vec{E}_g \approx -\partial \vec{H}_g/\partial t, etc. \tag{7.26}$$

$$\vec{E}_g = -\nabla\phi - \partial\vec{A}/\partial t, \quad \vec{H}_g = \nabla \times \vec{A} \tag{7.27}$$

$$\phi \approx -\frac{1}{2}(g_{00} + 1), \quad \vec{A}_i \approx g_{0i}. \tag{7.28}$$

It must be borne in mind that the fields E and H do not really represent the electromagnetic field, but bear a superficial resemblance. The subscripts g in Eqs. (7.26) and (7.27) bring out this distinction. Further using Eq. (7.27) in Eq. (7.24), we get

$$|\vec{H}| \approx \int \frac{\rho V}{r^2}\vec{r} \approx \frac{mV}{r^2}, \tag{7.29}$$

where the approximations indicate that these are order of magnitude equations as well as

$$|\vec{E}| = \frac{mV^2}{r^2}, \tag{7.30}$$

where V is the speed.

These approximations are consistent if in (7.29) and (7.30) the distance $r >>$ than the Compton scale.

Recall that by Heisenberg's principle:

$$mVr \approx h$$

whence we get

$$|\vec{H}| \sim \frac{h}{r^3}, |\vec{E}| \sim \frac{hV}{r^3}. \tag{7.31}$$

An interesting observation at this stage would be that (7.31) has no particle mass. We will now try to look at this new force in greater detail.

Indeed from Taylor [19], we can see that the Kerr–Newman metric gives a suitable description of the electron. Furthermore, here is an interesting feature, the anomalous gyromagnetic ratio, $g = 2$ is also recovered. Using a different mode of reasoning, Nottale [22] also reaches the same conclusion. It must be stressed that interestingly enough the Kerr–Newman field has additional electric and magnetic terms (Cf. [23]), both of $\mathcal{O}(\frac{1}{r^3})$. Remarkably this may be obtained from (7.31).

A pertinent consideration at this juncture would be to investigate the possible existence of such a force. To sum up we need mass independence, and spin dependence with a very short range. Maybe the hypothesized and mysterious $B_{(3)}$ [24] short-range force proposed by Evans would fill the bill.

Unexpectedly, if the force is carried by a "massive" particle, that is, within a massive vector field, we can get back (7.30) and (7.31) [25].

Finally, we must bear in mind that the resemblance of equations like (7.26), (7.27), and (7.28) with those of electromagnetism is but accidental. This can be treated as an unexpected coincidence. In the Kerr–Newman metric formulation, there is a similarity to electromagnetism. But this, however, is not fortuitous, as it is the metric that gives rise to both electromagnetism and gravitation (Cf. also Refs. [2, 26, 27]).

7.3 Observational Conclusions

A few years ago, some researchers from Hungary asserted that there is a new particle—called X17 which they discovered. It was their hypothesis that X17 would describe a fifth force of nature. Although there is a lot of scepticism about the existence of this particle.

The scientific community is looking forward to the ratification of such a discovery. In the same vein, the Jefferson Laboratory performing the DarkLight experiment is trying to establish the existence of dark photons with masses in the range of 10–100 MeV. They claim that this discovery has been made. Some physicists feels that this discovery strengthens the belief that there is indeed new force. It must be mentioned here however that some of the so-called new particles discovered have later not shown up. In some experiments, some excess decay signals were detected. This hinted of the existence of a new and very weak particle. If indeed there is such a particle its estimated mass would be roughly 50 times smaller than that of the proton. And due to its properties, it would be a boson. The findings were published in the Physical Review Letters [4, 5, 28].

The Large Hadron Collider (LHC) physicists, [6] in March 2021, reported hints of a new physics, viz. a possible fifth force. The LHC was referring to its beauty or bottom quark observations. A paper was published by this group in March 2021 based on results from the LHCb experiment. Using ultra-high-energy collisions the LHC studied the decay of beauty quarks. They discovered that the rate at which beauty quarks decayed into electrons and muons was different. This result was unexpected. Because one expects these rates to be the same. That is because a beauty quark employs the weak force to decay into electrons or muons with equal rates. According to the LHC team, muon decay was taking place at about 85% as that of the electron decay. They reasoned that this would require a new force of nature which would act on electrons and muons differently. This was the reason the team felt that there was a difference in the decay of beauty quarks. Of course this decay result had a statistical "three σ" to its credit. Generally a "five σ" would be required for certainty. Two other decays were also studied: one where the beauty quarks were paired with "down" quarks and another where they were also paired with "up" quarks. This time, muon decays happened around 70% as often as the electron decays and the result was a two σ result. This could be leading to a major discovery, but more experimental data is needed. There are also other experiments at the LHC, and in addition there is the Belle 2 experiment in Japan, which hopes to uncover more such results.

Quarks

Let us talk now about a quark–anti-quark pair which is commonly referred to as quarkonium. A special case of this would be charmonium, [29]. The spectrum of this quark–anti-quark pair has been calculated with a particular potential of the type $A/r + Br$, where A and B are constants. If we insert a $1/r^2$ term in the potential or $A/r + Br + C/r^2$ (from the Kerr–Newman metric), this maybe treated as a perturbation to the charmonium energy level. This perturbation leads to shifts in the energy levels. For example, let us consider a theoretical cavity without any of the known forces, then the wobble of the muon would still be detected.

In passing we would like to point out that the force could be of the type $S'U(2)$ [19], where in $S'U(2)$ the prime denotes that this is not the usual weak interaction.

Finally, it would seem that Fermilab has attained a 4.2 σ level of confidence [30] in detecting a new force.

References

1. B.G. Sidharth, *The Universe of Fluctuations* (Springer, 2005)
2. B.G. Sidharth, *Chaotic Universe* (Nova Science, New York, 2001)
3. S. Perlmutter, Discovery of a supernova explosion at half the age of the Universe. Nature **391**(6662), 51–54 (1998)
4. E. Cartlidge, Has a Hungarian physics lab found a fifth force of nature? Nature 1–2 (2016)
5. J. Krasznahorkay et al., Observation of anomalous internal pair creation in be: a possible indication of a light, neutral boson. Phys. Rev. Lett. **116**(4), 042501 (2016). (Jan)
6. R. Aaij, A.S.W. Abdelmotteleb, C.A. Beteta, Abudin é. n. F., et al. Tests of lepton universality using $b^0 \to k_s^0 \ell^+ \ell^-$ and $b^+ \to k^{*+} \ell^+ \ell^-$ decays (2021). arXiv:2110.09501
7. C.W. Misner, K.S. Thorne, *and J* (Freeman, A. Wheeler. Gravitation, 2000)
8. W. Greiner et al., textitRelativistic Quantum Mechanics, vol. 2 (Springer, 2000)

9. J.D. Bjorken, S.D. Drell, *Relativistic Quantum Mechanics* (Mcgraw-Hill College, 1964)
10. H. Feshbach, F. Villars, Elementary relativistic wave mechanics of spin 0 and spin 1/2 particles. Rev. Mod. Phys. **30**(1), 24–25 (1958)
11. B.G. Sidharth, Ultra high energy behaviour (2011). arXiv:1103.1496
12. B.G. Sidharth, *The Thermodynamic Universe* (World Scientific, Singapore, 2008)
13. G. Joos, I.M. Freeman, *Theoretical Physics* (Courier Corporation, 2013)
14. R. Penrose, C.J. Isham, *Quantum Concepts in Space and Time* (Oxford science publications, Clarendon Press, 1986)
15. B.G. Sidharth, Complexified Spacetime. Found. Phys. Lett. **16**(1), 91–97 (2003)
16. B.G. Sidharth, A brief note on "extension, spin and noncommutativity". Found. Phys. Lett. **15**(5), 501–506 (2002)
17. S.S. Schweber, *An Introduction to Relativistic Quantum Field Theory* (Courier Corporation, 2011)
18. T.D. Newton, E.P. Wigner, Localized states for elementary systems. Rev. Mod. Phys. **21**, 400–406 (1949). (Jul)
19. J.C. Taylor, *Gauge theories of Weak Interactions* (Cambridge, 1976)
20. B.G. Sidharth, Brief Report. New. Adv. Phys. **5**(1), 49–50 (2011)
21. B. Mashhoon, H.J. Paik, C.M. Will, Detection of the gravitomagnetic field using an orbiting superconducting gravity gradiometer. theoretical principles. Phys. Rev. D **39**(10), 2825 (1989)
22. L. Nottale, Scale Relativity and gauge invariance. Chaos, Solitons & Fractals **12**(9), 1577–1583 (2001)
23. E.T. Newman, Maxwell's equations and complex Minkowski space. J. Math. Phys. **14**(1), 102–103 (1973)
24. M.W. Evans, Origin, Observation and Consequences of the B" field, in *The Present Status of the Quantum Theory of Light: Proceedings of a Symposium in Honour of Jean-Pierre Vigier* (Springer Netherlands, 1997), pp. 117–125
25. C. Itzykson, J.B. Zuber, *Introduction to Quantum Field Theory* (McGraw-Hill International Book Company, 1980)
26. B.G. Sidharth, A reconciliation of Electromagnetism and Gravitation. Annales de la Fondation Louis de Broglie **27**(2), 333 (2002)
27. B.G. Sidharth, Gravitation and electromagnetism. Nuovo Cimento B Serie **116**(6), 735 (2001)
28. J.L. Feng, B. Fornal, I. Galon, S. Gardner, J. Smolinsky, T. Tait, P. Tanedo, Protophobic fifth-force interpretation of the observed anomaly in be 8 nuclear transitions. Phys. Rev. Lett. **117**(7), 071803 (2016)
29. B.H. Bransden, C.J. Joachain, *Introduction to Quantum Mechanics* (Longman Publishing Group, 1989)
30. B. Abi, T. Albahri, S. Al Kilani, et al., Measurement of the positive muon anomalous magnetic moment to 0.46 ppm. Phys. Rev. Lett. **126**, 141801 (2021)

Chapter 8
A Random Walk Through Miscellany

8.1 Fission with a Difference

A possible alternative route to releasing fission energy is considered here. Some recent work by the scientists Cruz Chu et al. [1] on this concept gives a clue for some further work. These physicists have done work which yielded nearly monochromatic radiation in the X-ray region.

The starting point is, from a relativistic point of view, with the Lorentz transformation:

$$x = \gamma(x' - vt), \quad \gamma = (1 - v^2/c^2)^{-1/2}. \tag{8.1}$$

As is known, for a collection of relativistic particles, the mass centres lead to a two-dimensional disc $\perp$ to the vector $\vec{L}$ denoting the angular momentum and with radius (Ref. [2]):

$$r = \frac{L}{mc}. \tag{8.2}$$

When the system has positive energy, it has a radius $> r$, while at distances $\mathcal{O}(r)$ we begin to encounter negative energies.

If we consider the system to be a particle of spin or angular momentum $L = \frac{\hbar}{2}$, then Eq. (8.2) gives $r = \frac{\hbar}{2mc}$. That leads us to the Compton wavelength region. Further, it must be borne in mind that the disc of mass centres is two dimensional.

Furthermore, we note that cf. Ref. [3]), if a spin half particle is represented by a Gaussian wave packet, then we come across negative energies at the Compton wavelength. Thus a particle can indeed be treated as a spherical shell of relativistic transient sub constituents or what may be called "particles". Indeed, this is another description of Dirac's *zitterbewegung* or rapid oscillation.

This is reminiscent of Dirac's shell or membrane model of the electron [4–6].

Outside this Compton region we have the usual Minkowski spacetime. But as we near the Compton wavelength region we come across a region where the space axis apparently transforms to a complex plane. This has been elaborated in detail by the

© The Author(s), under exclusive license to Springer Nature Singapore Pte Ltd. 2025
B. G. Sidharth, *The Dark Energy Paradigm*,
https://doi.org/10.1007/978-981-96-3745-4_8

author, using the Feshbach formalism [7] which leads to the double Wiener process. Consider the following system:

$$i\hbar\frac{\partial\phi}{\partial t} = \frac{1}{2m}\left(\frac{\hbar}{i\nabla} - \frac{e\mathbf{A}}{c}\right)^2 (\phi + \chi) + (e\phi + mc^2)\phi$$

$$i\hbar\frac{\partial\phi}{\partial t} = -\frac{1}{2m}\left(\frac{\hbar}{i\nabla} - \frac{e\mathbf{A}}{c}\right)^2 (\phi + \chi) + (e\phi - mc^2)\phi \tag{8.3}$$

Reference [8]. The merit of this formalism is that it yields a particle interpretation to the usual wave formulation (see [7] for further details). It needs to be pointed out that the advantage of the Feshbach–Villars' formalism is that we can now work with a particle interpretation.

As we have seen the Compton scale is encountered repeatedly. Wigner [9] pointed out the remarkable universality of the Compton scale.

From this characterization, it is clear that if an elementary particle is bombarded with very high-frequency radiation of the order of the Compton frequency, such an elementary particle would disintegrate, yielding its energy. The situation here is that the bell curve becomes compressed enough to be nearly a straight line or a spike, almost (see [10, 11]). It is this sharp spike under which the particle disintegrates giving up its mass as energy.

In quantum theory, monochromatic waves are an idealization. In the sense that what we actually have are wave packets with different energies [12]. But somehow is it possible to obtain a pure or nearly pure frequency? This of course is a problem of experimental technology. Let us examine the theoretical aspect. To see this, let us start with the Schrodinger equation:[12]

$$\frac{d^2\psi}{dx^2} + \frac{p^2}{\hbar^2}\psi = 0,$$

where

$$p = \sqrt{2m[E - V(x)]}.$$

This leads to

$$\phi(x)e^{\pm\frac{i}{\hbar}\int^x p(x)dx}, \tag{8.4}$$

where it is evident that there is a wave packet with different values of p or effectively frequencies. However, a hypothetical wave function like $\psi' = e^{ikx-pt}$ would be an idealization and would be monochromatic. Is this achievable? Some evidence is there due to the work of Cruz-Chu and co-workers [1] who have experimental work where single particle X-ray diffraction patterns could be analysed using a machine learning algorithm.

In recent years, there has been some further progress in this direction [10, 11, 13]. A purely monochromatic signal would come in handy for communication theory as well. This is because, this causes an effective increase in the bandwidth [14]. An observation here is that if we split quarkonium particles, even greater energy maybe released. There is also an alternative way for this: Using $g = 2$, there is a sort of a precession and, if we could radiate the particles with resonant frequencies, the particle would break up. This again could be a problem of technically realizing it.

8.2   Ultra-High-Energy Decaying Fermions

In this section, we return to ultra-relativistic fermions and their behaviour, with new inputs from quantum gravity approaches [15–17]. This is extremely relevant as the Large Hadron Collider in Geneva has already attained 7 TeV and is on the verge of attaining full energy 14 TeV.

No accelerator to date has achieved such high energies. Bearing this in mind we analyse ultra-high-energy fermionic collisions.

8.2.1   The High-Energy Equation

It is known that at very high energies, we encounter negative energy solutions. This is because the set of positive-energy solutions of the Dirac or Klein–Gordon equations is not a complete set [7] and so cannot describe a particle localized in any sense. At usual energies we could apply the well-known Foldy–Wouthuysen transformation to recover a description in terms of positive energies alone or more precisely a description free of operators which mix negative-energy and positive-energy components of the wave function. This description also leads in the non-relativistic limit to the two-component Pauli equation [3]. All this is well known.

In the case of very high energies, it was shown several years ago by Cini and Toushek that we can modify the Foldy–Wouthuysen transformation and obtain a different description [19]. Let us examine this situation in greater detail [20].

The Cini–Toushek transformation can be written in the form

$$e^{\pm \iota s} = \frac{E + |p|}{2E} \pm \frac{\vec{\gamma} \cdot \vec{p}}{2E|p|} \cdot m, \tag{8.5}$$

where the symbols like E, p, and γ have their usual meaning. Under (8.5), it is well known that the Dirac equation takes on the form of the massless neutrino equation:

$$H\psi - \frac{\vec{\alpha} \cdot \vec{p}}{|p|} E(p)\psi.$$

In the above we use the following notation:

$$\alpha^k = \begin{pmatrix} 0 & \sigma^k \\ \sigma^k & 0 \end{pmatrix} \qquad \beta = \begin{pmatrix} I & 0 \\ 0 & -I \end{pmatrix} \tag{8.6}$$

$$\gamma^0 = \beta \tag{8.7}$$

$$\gamma^k = \beta\alpha^k \quad (k = 1, 2, 3), \tag{8.8}$$

where σ^k are the Pauli matrices and I is the 2×2 unit matrix.

We will also require the transformation of the γ_5 operator, which is given by

$$\gamma^5 = \gamma^0\gamma^1\gamma^2\gamma^3 = \iota \begin{pmatrix} I & 0 \\ 0 & -I \end{pmatrix}. \tag{8.9}$$

Using (8.5), the transformed matrix is given by

$$\Gamma_5 = e^{-\iota s}\gamma_5 e^{\iota s} = \left\{ \frac{E+p}{2E} + \frac{(\vec{\gamma} \cdot \vec{n})m}{2E} \right\} \gamma_5 \left\{ \frac{E+p}{2E} - \frac{(\vec{\gamma} \cdot \vec{n})m}{2E} \right\} \tag{8.10}$$

which finally reduces to

$$\Gamma_5 = \gamma_5 + \left(\frac{m}{E}\right)(\vec{\gamma} \cdot \vec{n})\gamma_5. \tag{8.11}$$

In the above $\vec{n}$ is the unit vector in the direction of the momentum vector. We can see from (8.11) that

$$\Gamma_5 = \gamma_5 \tag{8.12}$$

whenever m vanishes. This is of course the well-known two-component neutrino case where the wave function can be decomposed into the left-handed and right-handed neutrino wave functions. Let us use (8.11) to proceed along similar lines and write

$$\psi = \psi_1 + \psi_2, \tag{8.13}$$

where

$$\psi_1 = \frac{1}{2}(1 - \gamma_5)\psi \text{ and } \psi_2 = \frac{1}{2}(1 + \iota\gamma_5)\psi. \tag{8.14}$$

If (8.12) were to hold, as for the neutrinos, then (8.14) would be the decomposition in terms of the left-handed and right-handed wave functions. If the mass does not vanish, that is, (8.12) does not hold then we will have from (8.14)

$$\psi_1 = (1 + \frac{m}{E})(1 + \gamma_5)\psi - \frac{m}{E}\psi \equiv (1 + \frac{m}{E})\psi_L - \frac{m}{E}\psi \tag{8.15}$$

with a similar equation for ψ_2. Equations (8.13) and (8.15) show that if $\frac{m}{E}$ is much less than 1, that is, when the energy is much greater than the rest energy, then we have a nearly two-component neutrino-like situation. We could, for example, interpret (8.13) and (8.14) as a decomposition into the left- and right-handed wave functions where the particle, as can be seen from (8.15), nearly exhibits handedness. Or more specifically as can be seen from (8.15) the wave function has a large part that displays handedness and a small part which is the usual type of wave function. More generally we can write (8.15) as

$$\psi = \psi_{\hat{H}} + \omega\psi_\Delta, \tag{8.16}$$

where $\psi_{\hat{H}}$ is the handed part and the second term is a small correction.

It must be borne in mind that when the total energy is much greater than the rest energy (8.16) holds. One could hope to see the effects, hopefully in the LHC which as remarked has already reached the 7 TeV mark and is reaching the 14 TeV mark.

8.2.2 Possible Consequences at High Energies

Firstly, it must be observed that the above theory becomes relevant in view of the fact that the neutrino is now known to have a mass, though the mass values are not yet certain, unlike the mass differences. This is because equations like (8.14), (8.15), and (8.16) can now be applied to neutrinos. Apart from this, the above shows that fermions in general behave like "heavy" neutrinos at very high energies. In any case as can be seen, these equations imply that apart from a $\mathcal{O}(\frac{m}{E})$ correction, γ_5 gets multiplied effectively by a factor $(1 + \mathcal{O}(\frac{m}{E}))$ (Cf. (8.15)). This means that in the usual Salam–Weinberg theory, a typical interaction term gets multiplied by a factor $(1 + \mathcal{O}(\frac{m}{E}))$ [21]

$$2^{\frac{1}{2}}G_W \left\{ \bar{\nu}_\mu \gamma^\lambda \frac{1}{2}(1+\gamma_5)\nu_\mu \right\} \left\{ \bar{e}\gamma_\lambda \left[\frac{1}{2}(1+\gamma_5)c_L + \frac{1}{2}(1-\gamma_5)c_R \right] e \right\} \left(1 + \mathcal{O}\left(\frac{m}{E}\right) \right). \tag{8.17}$$

That is, c_L and c_R are also multiplied by a similar small deviation from unity to become c'_L, c'_R. This in turn implies that the differential cross section now becomes in terms of the fermion recoil energy E'

$$\frac{d\sigma}{dE'_0} = [G_w^2/(2\pi m_e E_\nu^2)][|c'_l|^2(p\cdot q)^2 + |c'_R|^2(p'\cdot q)^2$$

$$+ \frac{1}{2}(c'^*_R c'_L + c'^*_L c'_R)m_0^2 q\cdot q']. \tag{8.18}$$

In any case the use of Γ_5 given by (8.11) instead of γ_5 would mean that a decay process would be asymmetrical in the angular distribution of the type $(1 + P\cos\Theta)$ where P is the average polarization.

The point is that fermions at such high energies would show handedness in accordance with (8.15) or (8.16). The possibility of CP violation in ultra-high-energy cosmic rays has been discussed by Collady and others [22]. In any case, these effects would have been present in the early universe.

Sudarshan et al. [23] use a similar analysis to get positive and negative energy operators $x_\pm$ for position and similar momentum operators, but interestingly they show that the x- and y-components do not commute. Sudarshan and co-workers introduced a sub- or superscript D and E for the Dirac and extreme relativistic (that is, Cini–Toushek type) representations. Then they deduced

$$[x_\pm, y_\pm] = \left(\frac{\imath p_2}{2p^3}\gamma_5\Lambda_\pm E\right)_{E\,repres.}$$

$$= \left(\pm\frac{\sigma_2}{2\imath p^2}\Lambda_\pm D\right)_{D\,repres.}, \tag{8.19}$$

where Λ is a projection operator which is given by

$$\Lambda_\pm = \frac{1}{2}(1 \pm H/E)$$

in the considered representation. This matter was investigated earlier by Newton and Wigner too [9] from a slightly different angle. Some years ago the author revisited this aspect from yet another point of view [24] and showed that this noncommutativity which is exhibited by (8.19) is related to spin and extension. The noncommutative nature of spacetime has been a matter of renewed interest in recent years particularly in quantum gravity approaches. At very high energies, it has been argued that [8] there is a minimum fuzzy interval, symptomatic of a noncommutative spacetime, so the usual energy–momentum relation gets modified and becomes

$$E^2 = p^2 + m^2 + \alpha l^2 p^4 \tag{8.20}$$

the so-called Snyder–Sidharth dispersion relation [25–27]. Using (8.20) it is possible to deduce the ultra-relativistic Dirac equation [28]:

$$(D + \beta l p^2 \gamma^5)\psi = 0 \tag{8.21}$$

$\beta = \sqrt{\alpha}$. In (8.21), D is the usual Dirac operator above while the extra term appears due to the new dispersion relation (8.20). We can see from (8.21) that the dispersion relation now becomes non-Hermitian and takes on an extra term (α) being positive (Cf. Ref. [18]):

$$H = M - \imath N, \tag{8.22}$$

where M is the usual Hamiltonian and N is now Hermitian (Cf. [29]), that is, M and N are real. This indicates a decay. With the modified Dirac equation (8.21) in place of the usual Dirac equation, we can now treat the two states considered above, viz.:

$$\psi_L, \psi_R$$

as forming a two-state system in this sub-space of the Hilbert space of all states where the two components decay at different rates, in general as we will see below. The theory of such two-state systems is well known [30]. In fact the two states would now be given by

$$\psi_{L,R}(t) = e^{\iota M t} \cdot e^{-N t} \psi_{L,R}(0), \tag{8.23}$$

where the left side refers to the state of time t and the right side is the wave function at the time $t = 0$ (Cf. also [31]). We can write the Hamiltonian (8.22) above for the two state as

$$H_{eff} = \begin{pmatrix} H_{11} & H_{12} \\ H_{21} & H_{22} \end{pmatrix} = M - \iota N = \begin{pmatrix} M_{11} & M_{12} \\ M_{21} & M_{22} \end{pmatrix} - \iota \begin{pmatrix} N_{11} & N_{12} \\ N_{21} & N_{22} \end{pmatrix}$$

where, by virtue of the imaginary term ι, both M and N are Hermitian. An additional constraint, namely, $H_{11} = H_{22}$, comes from the CPT theorem. Let us continue with the two-state analysis.

The evolution equation (in this sub-space)

$$H|\psi\rangle = \iota \frac{d}{dt}|\psi\rangle$$

yields the usual solution

$$|\psi_{H,L}\rangle(t) = exp[-\iota H_{H,L}]|\psi_{H,L}\rangle(0),$$

where $H_{H,L}$ denotes the eigenvalues of H, which are under the assumption of CPT symmetry given, as is well known, by

$$H_{H,L} = H_{11} \pm \sqrt{H_{12} H_{21}}$$

and $|\psi_{H,L}\rangle$ are eigenstates of the form:

$$|\psi_{H,L}\rangle = p|\psi^0\rangle \mp q|\psi^0\rangle$$

with

$$\frac{q}{p} = -\frac{H_H - H_L}{2H_{12}}.$$

Rewriting the time-dependent solution using $H_{H,L} = M_{H,L} - \iota N_{H,L}$ with real M and N, we get

$$|\psi_{H,L}\rangle(t) = exp\left[-N_{H,L}\right] exp[-\iota M_{H,L}](t)|\psi_{H,L}\rangle(0).$$

This represents two fermions (one perhaps heavier with mass M_H, one lighter with mass M_L), decaying with (generally different) decay constants $N_{H,L}$. The mean mass $M = \frac{1}{2}(M_H + M_L)$ and $\Delta M = M_H - M_L$. It has been pointed out that equations like (8.14), (8.15), or (8.16) applied to neutrinos which are massless suggest one (or more) neutrinos.

In any case, this analysis is true for fermions in general and one would expect handedness and even decomposition at very high energies. One could look at it in the following way. The extra term in the new Hamiltonian (8.20), the modified Dirac equation (8.21), and the non-Hermitian Hamiltonian (8.22) split the state, much like the introduction of a magnetic field leading to the Zeeman splitting.

## 8.3	A Zeeman-Like Effect

Consider two monochromatic waves, which can reinforce each other or destructively come to nothing. In case of constructive interference, we can consider it from the point of view of fusion. For this we use the beats interpretation of quantum mechanics. While the waves which are constructively lumped together can be considered as fused particles. We have fusion, otherwise it is just energy. We could even construct a wave packet. The question to be posed is that in general do we get fusion and energy? All this can be seen within the framework of two recent technological advances. One is the isolation of monochromatic waves [1]. The other is the realization of ultra short-range interactions by using laser cooling techniques [32]. We have already touched upon graphene or any other two-dimensional honeycomb-like structures acting as a test bed for high-energy physics. Now let us suppose that a stream of high-energy particles, for example, neutrinos pass across this layer, then this volley would broadly consist of a mode of particles, let us say luminal and outliers, some of which may be superluminal. This has been argued elsewhere and what is equally important, this could explain the anomaly of the 1987 supernova emissions.

Earlier, the author had derived an expression for a grand unified Lagrangian (presented in Frontiers of Fundamental Physics 15, Orihuela [33]) (as seen earlier). Further, the author had shown that based on noncommutativity it is possible to reconcile electromagnetism and gravitation [34]. Finally, more recently the author along with the late Larissa Lapershvili, Nielsen et al., had reconciled weak interaction with gravitation [35]. This shows that, in this manner, it is possible to reconcile all four fundamental interactions in a unified manner.

8.4 Epilogue: The Gravitation Frontier

The ultimate unification achieved? This eluded Einstein while for a century physicists have been trying to obtain a unified description of gravitation and electromagnetism. Finally, the great Wolfgang Pauli declared "do not try to combine what Nature had meant to be separate". There were attempts over the decades by several physicists, for example, Herman Weyl tried this combination, though Einstein rejected it, quite rightly, that gravitation was put in by hand. The author has tried this unification, and this is described in his book Thermodynamic Universe (World Scientific) [8]. Here the author brings out the fact that every point in space is bispinorial. This is because, if we consider the 4×4 Dirac matrices, they are really made up of two 2-spinors (as mentioned earlier) with ϕ representing the positive-energy solutions and χ the negative-energy solutions. These have opposite properties under space reflections. This leads to, as explained in detail in the above reference, a formula that is mathematically identical to the Weyl formulation, except that Einstein's objection is no longer valid because, now, it is the micro-property of spacetime that brings out a combined description. This is the desired unification.

References

1. E.R. Cruz-Chú, A. Hosseinizadeh, G. Mashayekhi, R. Fung, A. Ourmazd, P. Schwander, Selecting XFEL single-particle snapshots by geometric machine learning. Struct. Dyn. **8**(1), 014701 (2021)
2. C. Moller, *The Theory of Relativity* (Oxford, 1952)
3. J.D. Bjorken, S.D. Drell, *Relativistic Quantum Mechanics* (Mcgraw-Hill College, 1964)
4. J.D. Bjorken, S.D. Drell, *Relativistic Quantum Mechanics* (Mcgraw-Hill College, 1964)
5. A.O. Barut, M. Pavsic, Dirac's Shell Model of the electron and the general theory of moving relativistic charged membranes. ICTP Preprint IC/92/399
6. A.O. Barut, M. Pavsic, The spinning minimal surfaces without the Grassmann variables. ICTP Preprint IC/88/2
7. H. Feshbach, F. Villars, Elementary relativistic wave mechanics of spin 0 and spin 1/2 particles. Rev. Mod. Phys. **30**(1), 24–25 (1958)
8. B.G. Sidharth, *The Thermodynamic Universe* (World Scientific, Singapore, 2008)
9. T.D. Newton, E.P. Wigner, Localized states for elementary systems. Rev. Mod. Phys. **21**, 400–406 (1949). (Jul)
10. M.V. Berry, M.R. Dennis, Natural superoscillations in monochromatic waves in D dimensions. J. Phys. A: Math. Theor. **42**(2), 022003 (2008)
11. M. Kostylev, G. Gubbiotti, G. Carlotti, G. Socino, S. Tacchi, C. Wang, N. Singh, A. Adeyeye, R.L. Stamps, Propagating volume and localized spin wave modes on a lattice of circular magnetic antidots. J. Appl. Phys. **103**(7), 07C507 (2008)
12. J.L. Powell, B. Crasemann, *Quantum Mechanics* (Courier Dover Publications, 2015)
13. J. Rodriguez, Parametric x-ray methods use 2D heterostructures to generate compact, tunable X-ray sources. Scilight **2021**(28), 281103 (2021)
14. B.G. Sidharth, Science scapes. New Adv. Phys. **2**(2), 2–5 (2021)
15. B. G. Sidharth, Ultra high energy behavior (2011). arXiv preprint arXiv:1103.1496
16. B. G. Sidharth, Ultra high energy fermions (2015). arXiv:1008.2491
17. B. G. Sidharth, Ultra high energy particles (2011). arxiv:1107.4459

18. B.G. Sidharth, High energy P-P collisions. Int. J. Mod. Phys. E **19**(10), 2023–2031 (2010)
19. M. Cini, B. Touschek, The relativistic limit of the theory of spin 1/2 particles. Il Nuovo Cimento **7**(3), 422–423 (1958)
20. S.S. Schweber, *An introduction to relativistic Quantum Field Theory* (Courier Corporation, 2011)
21. J.C. Taylor, *Gauge theories of Weak Interactions* (Cambridge, 1976)
22. D. Colladay, New implications of Lorentz violation. Int. J. Mod. Phys. A **20**(06), 1260–1267 (2005)
23. S.K. Bose, A. Gamba, E.C.G. Sudarshan, Representations of the Dirac equation. Phys. Rev. **113**(6), 1661 (1959)
24. B.G. Sidharth, A brief note on "extension, spin and noncommutativity". Found. Phys. Lett. **15**(5), 501–506 (2002)
25. B.G. Sidharth, Different routes to Lorentz symmetry violations. Found. Phys. **38**(1), 89–95 (2008)
26. B.G. Sidharth, Noncommutative spacetime, mass generation, and other effects. Int. J. Mod. Phys. E **19**(01), 79–90 (2010)
27. L.A. Glinka, On some consequences of the Snyder-Sidharth deformation of special Relativity. Apeiron: Stud. Infinite Nat. **16**(2), 147 (2009)
28. B.G. Sidharth, A note on noncommutativity and mass generation. Int. J. Mod. Phys. E **14**(06), 923–925 (2005)
29. B.G. Sidharth, The mass of the neutrinos. Int. J. Mod. Phys. E **19**(11), 2285–2292 (2010)
30. R.P. Feynman, R.B. Leighton, M. Sands, The Feynman lectures on physics; vol. i. Am. J. Phys. **33**(9), 750–752 (1965)
31. L. Widhalm et al., Leptonic and semileptonic d and d_s decays at b-factories (2007). arXiv:0710.0420
32. A.A. Geraci, B.P. Scott, J. Kitching, Short-range force detection using optically cooled levitated microspheres. Phys. Rev. Lett. **105**(10), 101101 (2010). (Aug)
33. Going beyond the Standard model. Fundamental Physics and Physics Education Research. (Springer, Cham), pp. 17–20
34. B.G. Sidharth, A reconciliation of Electromagnetism and Gravitation. Annales de la Fondation Louis de Broglie **27**(2), 333 (2002)
35. B.G. Sidharth, L.V. Lapershvili, H.B Nielsen, et al., Gravi-weak unification and the black-hole-hedgehog's solution with magnetic field contribution. Int. J. Mod. Phys. A **33**(32), 1850188 (2018)

Made in the USA
Monee, IL
07 July 2026

56553664R00098